Communications in Computer and Information Science 1062

Commenced Publication in 2007
Founding and Former Series Editors:
Phoebe Chen, Alfredo Cuzzocrea, Xiaoyong Du, Orhun Kara, Ting Liu,
Krishna M. Sivalingam, Dominik Ślęzak, Takashi Washio, and Xiaokang Yang

More information about this series at http://www.springer.com/series/7899

Gabriele Anderst-Kotsis ·
A Min Tjoa · Ismail Khalil et al. (Eds.)

Database and Expert Systems Applications

DEXA 2019 International Workshops
BIOKDD, IWCFS, MLKgraphs and TIR
Linz, Austria, August 26–29, 2019
Proceedings

 Springer

Editors
Gabriele Anderst-Kotsis
Institute of Telecooperation
Johannes Kepler University of Linz
Linz, Oberösterreich, Austria

A Min Tjoa ⓘ
Software Competence Center Hagenberg
Hagenberg, Austria

Ismail Khalil
Institute of Telecooperation
Johannes Kepler University of Linz
Linz, Oberösterreich, Austria

Additional Editors *see next page*

ISSN 1865-0929   ISSN 1865-0937   (electronic)
Communications in Computer and Information Science
ISBN 978-3-030-27683-6   ISBN 978-3-030-27684-3   (eBook)
https://doi.org/10.1007/978-3-030-27684-3

This Springer imprint is published by the registered company Springer Nature Switzerland AG
The registered company address is: Gewerbestrasse 11, 6330 Cham, Switzerland

Additional Editors

BIOKDD Workshop
Elloumi Mourad
LaTICE, University of Tunis
Tunisia
mourad.elloumi@gmail.com

IWCFS Workshop
Atif Mashkoor
Software Competence Center Hagenberg
Austria
Atif.Mashkoor@scch.at

Sametinger Johannes
Johannes Kepler University Linz
Austria
Johannes.Sametinger@jku.at

Larrucea Xabier
Tecnalia Research & Innovation
Spain
xabier.larrucea@tecnalia.com

MLKgraphs Workshop
Fensel Anna
University of Innsbruck
Austria
anna.fensel@sti2.at

Martinez-Gil Jorge
Software Competence Center Hagenberg
Austria
jorge.Martinez-Gil@scch.at

Moser Bernhard
Software Competence Center Hagenberg
Austria
bernhard.moser@scch.at

TIR Workshop
Seifert Christin
University of Twente
The Netherlands
c.seifert@utwente.nl

Stein Benno
Bauhaus-Universität Weimar
Germany
benno.stein@uni-weimar.de

Granitzer Michael
University of Passau
Germany
michael.granitzer@uni-passau.de

Preface

The Database and Expert Systems Applications (DEXA) workshops are a platform for the exchange of ideas, experiences, and opinions among scientists and practitioners – those who are defining the requirements for future systems in the areas of database and artificial technologies.

This year DEXA featured four international workshops:

- BIOKDD 2019, the 10th International Workshop on Biological Knowledge Discovery from Data
- IWCFS 2019, the Third International Workshop on Cyber-Security and Functional Safety in Cyber-Physical Systems
- MLKgraphs2019, the First International Workshop on Machine Learning and Knowledge Graphs
- TIR 2019, the 16th International Workshop on Technologies for Information Retrieval

DEXA workshops papers included papers that focus mainly on very specialized topics on the applications of database and expert systems technology.

We would like to thank all workshop chairs and Program Committee members for their excellent work, namely, Mourad Elloumi, chair of the BIOKDD workshop, Atif Mashkoor, Johannes Sametinger, and Xabier Larrucea, co-chairs of the IWCFS workshop, Anna Fensel, Jorge Martinez-Gil, and Bernhard Moser, co-chairs of the MLKgraphs workshop, and Christin Seifert, Benno Stein, and Michael Granitzer, co-chairs of the TIR workshop.

DEXA 2019 was the 30th in the series of annual scientific platform on database and expert systems applications after Vienna, Berlin, Valencia, Prague, Athens, London, Zurich, Toulouse, Vienna, Florence, Greenwich, Munich, Aix en Provence, Prague, Zaragoza, Copenhagen, Krakow, Regensburg, Turin, Linz, Bilbao, Toulouse, Vienna, Prague, Munich, Valencia, Porto, Lyon, and Regensburg.

This year DEXA took place at the Johannes Kepler University of Linz, Austria.

We would like to express our thanks to all institutions actively supporting this event, namely:

- Johannes Kepler University Linz (JKU)
- Software Competence Center Hagenberg (SCCH)
- International Organization for Information Integration and Web-Based Applications and Services (@WAS)

Finally, we hope that all the participants of the DEXA 2019 workshops enjoyed the program that we put together.

August 2019

Gabriele Anderst-Kotsis
A Min Tjoa
Ismail Khalil

Organization

Steering Committee

Gabriele Anderst-Kotsis   Johannes Kepler University Linz, Austria
A Min Tjoa   Technical University of Vienna, Austria
Ismail Khalil   Johannes Kepler University Linz, Austria

BIOKDD 2019 Chair

Mourad Elloumi   LaTICE, University of Tunis, Tunisia

BIOKDD 2019 Program Committee and Reviewers

Jamal Al Qundus   Freie Universität Berlin, Germany
Matteo Comin   University of Padova, Italy
Bhaskar DasGupta   University of Illinois at Chicago (UIC), USA
Adrien Goëffon   University of Angers, France
Robert Harrison   Georgia State University, USA
Daisuke Kihara   Purdue University, USA
Giuseppe Lancia   University of Udine, Italy
Dominique Lavenier   IRISA, France
Vladimir Makarenkov   The Universite du Quebec Montreal, Canada
Solon P. Pissis   Centrum Wiskunde & Informatica (CWI),
                             The Netherlands
Davide Verzotto   University of Pisa, Italy
Emanuel Weitschek   Uninettuno International University, Italy
Malik Yousef   Zefat Academic College, Israel
Farhana Zulkernine   Queen's University, Canada

IWCFS 2019 Chairs

Atif Mashkoor   Software Competence Center Hagenberg, Austria
Johannes Sametinger   Johannes Kepler University Linz, Austria
Xabier Larrucea   Tecnalia Research & Innovation, Spain

IWCFS 2019 Program Committee and Reviewers

Yamine Ait Ameur   INPT-ENSEEIHT/IRIT, France
Egyed Alexander   JKU Linz, Austria
Gargantini Angelo   University of Bergamo, Italy
Paolo Arcaini   National Institute of Informatics, Japan
Richard Banach   University of Manchester, UK

Ladjel Bellatreche	LISI/ENSMA, France
Ladjel Bellatreche	ENSMA, France
Miklos Biro	Software Competence Center Hagenberg GmbH, Austria
Jorge Cuellar	University of Passau, Germany
Osman Hasan	National University of Science and Technology, Pakistan
Jean-Pierre Jacquot	University of Lorraine, France
Muhammad Khan	University of Surrey, UK
Xabier Larrucea	Tecnalia Research & Innovation, Spain
Atif Mashkoor	SCCH, Austria
Muhammad Muaaz	University of Agder, Norway
Singh Neeraj	IRIT, France
Aakarsh Rao	University of Arizona, USA
Martin Ochoa Ronderos	Universidad del Rosario, Colombia
Johannes Sametinger	Johannes Kepler University Linz, Austria
Remzi Seker	Embry-Riddle Aeronautical University, USA
Edgar Weippl	SBA, Austria

MLKgraphs 2019 Chairs

Anna Fensel	University of Innsbruck, Austria
Jorge Martinez-Gil	Software Competence Center Hagenberg, Austria
Bernhard Moser	Software Competence Center Hagenberg, Austria

MLKgraphs 2019 Program Committee and Reviewers

Alessandro Adamou	Insight Centre for Data Analytics, Ireland
Mohamed Ben Ellefi	Aix Marseille Universite, France
Jose Manuel Chaves-Gonzalez	University of Extremadura, Spain
Ioana Ciuciu	UBB Cluj, Romania
Lisa Ehrlinger	Johannes Kepler University Linz and Software Competence Center Hagenberg GmbH, Austria
Anna Fensel	University of Innsbruck, Austria
Isaac Lera	University of Balearic Island, Spain
Jorge Martinez-Gil	Software Competence Center Hagenberg, Austria
Ismael Navas-Delgado	University of Málaga, Spain
Mario Pichler	Software Competence Center Hagenberg GmbH, Austria
Marta Sabou	TU Wien, Austria
Harald Sack	FIZ Karlsruhe and Karlsruhe Institute of Technology, Germany
Iztok Savnik	University of Primorska, Slovenia
Francois Scharffe	Columbia University, USA
Marina Tropmann-Frick	Hamburg University of Applied Sciences, Germany

TIR 2019 Chairs

Christin Seifert   University of Twente, The Netherlands
Benno Stein   Bauhaus-Universität Weimar, Germany
Michael Granitzer   University of Passau, Germany

TIR Program Committee and Reviewers

Mikhail Alexandrov   Autonomous University of Barcelona, Spain
Milad Alshomary   Paderborn University, Germany
Alberto Barrón Cedeño   Qatar Computing Research Institute, Qatar
Alexander Bodarenko   Martin-Luther-Universität Halle-Wittenberg, Germany
Ingo Frommholz   University of Bedfordshire, UK
Christian Gütl   Technical University Graz, Austria
Chaker Jebari   CAS-IBRI-OMAN, Oman
Anna Kasprzik   ZBW-Leibniz Information Centre for Economics,
                          Germany
Nedim Lipka   Adobe Research, USA
Mihai Lupu   Vienna University of Technology, Austria
Harald Sack   FIZ Karlsruhe and Karlsruhe Institute of Technology,
                          Germany
Fatemeh Salehi Rizi   University of Passau, Germany
Marina Santini   RISE Research Institutes of Sweden, Sweden
Jörg Schlötterer   University of Passau, Germany
Christin Seifert   University of Twente, The Netherlands
Christian Wartena   Hochschule Hannover, Germany
Nils Witt   German National Library of Economics, Germany

Organizers

Contents

TIR

Biological Knowledge Discovery from Data

An Approach for LPF Table Computation

Supaporn Chairungsee[1](✉) and Thana Charuphanthuset[2]

[1] Walailak University, Nakhonsithammarat, Thailand
s.chairungsee@gmail.com
[2] Suratthani Rajabhat University, Suratthani, Thailand

Abstract. In this article, we introduce a new solution for the Longest Previous Factor (LPF) table computation. The LPF table is the table that stores the maximal length of factors re-occurring at each position of a string and this table is useful for text compression. The LPF table has the important role for computational biology, data compression and string algorithms. In this paper, we present an approach to compute the LPF table of a string from its suffix heap. The algorithm runs in linear time with linear memory space.

Keywords: Longest previous factor table · Data compression ·
Suffix heap · Text compression

1 Introduction

This study focuses on the problem of the Longest Previous Factor (LPF) table computation. The Longest Previous Factor (LPF) table is the table that stores, for each position of the string, the maximal length of factors occurring both there and at a previous position of the string. The LPF table was invented by Crochemore [1,3–6]. This table is useful for dictionary compression method and lossless data compression [2,8,11–13].

As an example, let us consider the LPF table of the string $y = $ abaabbbaa:

Position i	1	2	3	4	5	6	7	8	9
$y[i]$	a	b	a	a	b	b	b	a	a
LPF$[i]$	0	0	1	2	1	1	3	2	1

In 2008, the algorithm to compute the LPF array is presented [3]. This study applied suffix array to calculate the LPF table and this solution desires $O(n)$ time. After that, algorithms for the LPF table computation are proposed [4]. These solutions are linear-time algorithms. Next, in 2010, Crochemore et al. [5] presented efficient algorithms for two extensions of LPF table using suffix array. The running time of these solutions does not rely on the alphabet size and these approaches can compute the LPF table in linear time. In 2013, Crochemore et al. [6] invented a method for the LPF array computation. They presented the first time–space optimal algorithm and this method uses suffix array and the longest

© Springer Nature Switzerland AG 2019
G. Anderst-Kotsis et al. (Eds.): DEXA 2019 Workshops, CCIS 1062, pp. 3–7, 2019.
https://doi.org/10.1007/978-3-030-27684-3_1

common prefix array. In 2017, the algorithm to compute the longest previous factor table using suffix tray is proposed [10]. This method can compute the Longest Previous Factor table of the text, t, of length n in $O(n)$.

In this paper, we present an algorithm to compute the LPF table of a string from its suffix heap. The algorithm runs in linear time with linear memory space and it can be applied for text compression and string algorithms. The paper is divided into four parts. The next part, basic definition, presents useful concepts for our solution. Next, method, proposes our algorithm for the LPF table computation. Then, the summarize of our study is presented in conclusion.

2   Basic Definition

In this section, we recall the notions of the Longest Previous Factors (LPF) table, and a suffix heap that will be applied for our approach respectively.

2.1   LPF Table

The Longest Previous Factor (LPF) table is the table which stores the maximal length of the previous factor occurring at each position of the text [3–6]. The formal definition of the LPF table is described for $1 \leq i \leq n - 1$ by.

$$\text{LPF}[i] = \max \{k | y[i..i + k - 1] = y[j..j + k - 1], \text{ for some } 0 \leq j < i\}.$$

The concept of LPF table is close to the concept of the Longest Previous non-overlapping Factor (LPnF) table [7]. Since the LPnF table stores the maximal length of the previous factor occurring at each position of the text where the two occurrences do not overlap. While the LPF table stores the maximal length of the previous factor occurring at each position of the text where the two occurrences can be overlap. For instance, the LPF and LPnF table of the string $y = \mathsf{baabaabaaba}$ is presented as follows.

Position i	1	2	3	4	5	6	7	8	9	10	11
$y[i]$	b	a	a	b	a	a	b	a	a	b	a
LPF$[i]$	0	0	1	8	7	6	5	4	3	2	1
LPnF$[i]$	0	0	1	3	3	3	5	4	3	2	1

2.2   Suffix Heap

Let y refer to a string of length n, $y[1..n]$, and the string y ends with a special symbol $y[n] = \$$. Let $S\text{-}Heap(y)$ denote the suffix heap of the string y [9]. The states of the suffix heap are labelled 1 to n and the initial state is labelled 0. The suffix array, SA, of a string y is a lexicographically ordered array of the set $y_1, y_2, ..., y_n$ of all suffixes of y [2]. Let SA^{-1} be the inverse suffix array which is the inverse permutation. For all $i \in 1...n$, the value of $SA^{-1}[SA[i]] = i$. The path label of the state labelled i is a prefix of $y[SA[i]..n]$. As an example, the suffix heap of a string $y = \textsf{baabaabaaba\$}$ is presented in the following. This data structure includes the table of failure link, F, and the table of smallest position, sp. Let p refer to a state of $S\text{-}Heap(y)$ and state p is a class of factor of a string y. Let u be any string of a class of factor of a string y. Let $F(q)$ refer to the suffix target of state q as the equivalence of $s(u)$ [2]. Let attribute sp refer to the smallest position in the string y corresponds to each states of the suffix heap.

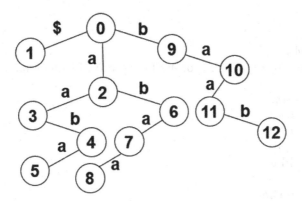

Fig. 1. Suffix heap of the string $y = \textsf{baabaabaaba\$}$.

The attributes F and sp for each states of the suffix heap that are displayed in Fig. 1 is presented in the table below. Let us consider the suffix target of state 4 in Fig. 1, we found that the suffix target of this state is state 8. If we consider the value of attribute sp for state 2, we found that the attribute sp for state 2 is 12.

State[q]	0	1	2	3	4	5	6	7	8	9	10	11	12
$F[q]$	0	0	0	2	6	7	9	2	11	0	2	3	4
$sp[q]$	0	12	2	2	2	2	3	3	3	1	1	1	1

3 Method

In this section, we present an approach to find the LPF table of a string y using its suffix heap. The pseudocode of our approach is presented as follows. At a given step q is the current state of the suffix heap and the initial state is defined as state 0. Let i be a position on y and l be the length of the current match. Let a be the character at position i of the string y and δ denote the transition function. The transition function of state q by character a is defined by $\delta(q, a)$. There are three conditions to be examined to extend the match. Firstly, we will examine whether the condition to extend the match by the letter $a = y[i]$ is $\delta(q, a)$ is defined. Next, we will investigate whether the condition to extend the match by the letter $a = y[i]$ is $\delta(F(q), a)$ is determined. After we will verify that the occurrence of $y[i-l..i-1]a$ in y at a smaller position than the position $i-l$.

If the result of the test is a mismatch, the failure link F is applied for shorten the match. Therefore, the value of LPF at position $[i - l]$ is defined as the length of the match.

Algorithm 1. LPF table with Suffix Heap $(y, n, S\text{-}Heap(y))$

```
1:  q ← 0
2:  l ← 0
3:  i ← 1
4:  repeat
5:      a ← y[i]
6:      while (δ(q, a) ≠ NULL or δ(F(q), a) ≠ NULL) and sp(δ(q, a)) + l < i
        do
7:          q ← δ(q, a)
8:          l ← l + 1
9:          i ← i + 1
10:         a ← y[i]
11:     end while
12:     LPF[i − l] ← l
13:     if q ≠ 0 then
14:         q ← F(q)
15:         l ← l − 1
16:     else
17:         i ← i + 1
18:     end if
19: until (i > n) and (l = 0)
20: return  LPF table
```

end

Theorem 1. *The algorithm LPF table computation with suffix heap computes the LPF table of a string of length n in time $O(n)$.*

Proof. The running time of the algorithm has been analyzed. First of all, we examine the suffix heap construction which includes the computation of attribute F and attribute sp. We found that this step, the computation of suffix heap, can be done in linear time [9]. Then, we consider the running time of the LPF computation. This step depends on the test to execute the next instruction and the test to increment of the expression $i - l$. As the values of these tests never decrease, the number of the tests are n. Then, the running time of the LPF table computation is $O(n)$. Therefore, the overall algorithm, the suffix heap construction and the LPF table computation, runs in $O(n)$ time.

4   Conclusion

We present an approach to find the Longest Previous Factor (LPF) table of the string. This table stores the maximal length of factors occurring at each position of a string. The LPF table is useful for data compression and text compression. Our solution computes the LPF table using the suffix heap of the string. The algorithm is a linear time with linear memory space.

References

1. Bell, T.C., Clearly, J.G., Witten, I.H.: Text Compression. Prentice Hall Inc., Upper Saddle River (1990)
2. Crochemore, M., Hancart, C., Lecroq, T.: Algorithms on Strings. Cambridge University Press, Cambridge (2007)
3. Crochemore, C., Ilie, L.: Computing longest previous factor in linear time and applications. Inf. Process. Lett. **106**(2), 75–80 (2008)
4. Crochemore, M., Ilie, L., Iliopoulos, C.S., Kubica, M., Rytter, W., Waleń, T.: LPF computation revisited. In: Fiala, J., Kratochvíl, J., Miller, M. (eds.) IWOCA 2009. LNCS, vol. 5874, pp. 158–169. Springer, Heidelberg (2009). https://doi.org/10.1007/978-3-642-10217-2_18
5. Crochemore, M., Iliopoulos, C.S., Kubica, M., Rytter, W., Waleń, T.: Efficient algorithms for two extensions of LPF table: the power of suffix arrays. In: van Leeuwen, J., Muscholl, A., Peleg, D., Pokorný, J., Rumpe, B. (eds.) SOFSEM 2010. LNCS, vol. 5901, pp. 296–307. Springer, Heidelberg (2010). https://doi.org/10.1007/978-3-642-11266-9_25
6. Crochemore, M., Ilie, L., Iliopoulos, C.S., Kubica, M., Rytter, W., Wale, T.: Computing the longest previous factor. Eur. J. Comb. **34**(1), 15–26 (2013)
7. Crochemore, C., Tischler, G.: Computing longest previous nonoverlapping factors. Inf. Process. Lett. **111**, 291–295 (2011)
8. Drozdek, A.: Data Structures and Algorithms in C++. Cengage Learning, Boston (2013)
9. Gagie, T., Hon, W.-K., Ku, T.-H.: New algorithms for position heaps. In: Fischer, J., Sanders, P. (eds.) CPM 2013. LNCS, vol. 7922, pp. 95–106. Springer, Heidelberg (2013). https://doi.org/10.1007/978-3-642-38905-4_11
10. Kongsen, J., Chairungsee, S.: Using suffix tray and longest previous factor for pattern searching. In: International Conference on Information Technology, Singapore, Singapore (2017)
11. Pu, I.M.: Fundamental Data Compression. A Butterworth-Heinemann, Oxford (2006)
12. Storer, J.A.: Data Compression: Methods and Theory. Computer Science Press, New York (1988)
13. Ziv, J., Lempel, A.: A universal algorithm for sequential data compression. IEEE Trans. Inf. Theory. **23**, 337–343 (1977)

Smart Persistence and Accessibility of Genomic and Clinical Data

Eleonora Cappelli[1], Emanuel Weitschek[2]([✉]), and Fabio Cumbo[3]

[1] Department of Engineering, Roma Tre University, 00146 Rome, Italy
`eleonora.cappelli@uniroma3.it`
[2] Department of Engineering, Uninettuno University, 00186 Rome, Italy
`emanuel@iasi.cnr.it`
[3] CIBIO Department, University of Trento, 38123 Trento, Italy
`fabio.cumbo@iasi.cnr.it`

Abstract. The continuous growth of experimental data generated by Next Generation Sequencing (NGS) machines has led to the adoption of advanced techniques to intelligently manage them. The advent of the Big Data era posed new challenges that led to the development of novel methods and tools, which were initially born to face with computational science problems, but which nowadays can be widely applied on biomedical data. In this work, we address two biomedical data management issues: (i) how to reduce the redundancy of genomic and clinical data, and (ii) how to make this big amount of data easily accessible. Firstly, we propose an approach to optimally organize genomic and clinical data by taking into account data redundancy and propose a method able to save as much space as possible by exploiting the power of no-SQL technologies. Then, we propose design principles for organizing biomedical data and make them easily accessible through the development of a collection of Application Programming Interfaces (APIs), in order to provide a flexible framework that we called OpenOmics. To prove the validity of our approach, we apply it on data extracted from The Genomic Data Commons repository. OpenOmics is free and open source for allowing everyone to extend the set of provided APIs with new features that may be able to answer specific biological questions. They are hosted on GitHub at the following address https://github.com/fabio-cumbo/open-omics-api/, publicly queryable at http://bioinformatics.iasi.cnr.it/openomics/api/routes, and their documentation is available at https://openomics.docs.apiary.io/.

Keywords: Biomedical data modeling ·
Biomedical data management · No-SQL · API ·
Genomic and clinical data

1 Introduction

During the last decade, many efforts have been done to deal with the problem of managing massive amounts of data produced by NGS technologies. A lot of

G. Anderst-Kotsis et al. (Eds.): DEXA 2019 Workshops, CCIS 1062, pp. 8–14, 2019.
https://doi.org/10.1007/978-3-030-27684-3_2

public and private repositories of genomic data have been created with the aim of spreading them [1–3]. Unfortunately, these databases often lack of standardization and of efficient storage models, resulting in waste of storage space. In this work, we face with the problem of modeling genomic and clinical data with a model that minimizes the amount of redundant information [4] (therefore also storage space), which is a big challenge that requires smart engineering solutions, by providing a set of Application Programming Interfaces (APIs) to guarantee an easy and efficient access to these data.

Recently, many efforts have been made towards a better management of genomic and clinical data through harmonization procedures for standardization and improved accessibility. Concrete examples of these methods are TCGA2BED [5] and OpenGDC [6], two software tools for the automatic extraction, extension, and standardization of public available genomic and clinical data from the The Cancer Genome Atlas (TCGA) portal [7], and the Genomic Data Commons (GDC) portal [8], respectively. The authors of these tools provide also open-access FTP repositories with standardized data in free-BED, which is widely used format in the bioinformatics community, available at http://bioinf.iasi.cnr.it/tcga2bed/ and at http://bioinf.iasi.cnr.it/opengdc/. In this work, we consider data from the OpenGDC repository that is constantly synchronized and updated with more recent GDC data of the TCGA program.

2   Methods

We took into account the free-BED data representation as starting point of our work. In particular, although these data are well organized following a clear conceptual data schema, they suffer of a data redundancy issue caused by the requirements imposed by the adopted standard.

Starting from the data representation described above, we extract all the free-BED standardized data from the OpenGDC public FTP repository and we model them to minimize their content redundancy. In particular, we focus on *Gene Expression Quantification* (GEQ) [9] and *Methylation Beta Value* (MBV) [10] data, which are the main experimental data types affected by the problem of redundant information.

Every GEQ experiment contains indeed the same information about the genes—genomic coordinates of the involved genes, the *ensembl gene id*, the *entrez gene id*, and *type*. Also MBV data contains redundant information consisting in the genomic coordinates of the involved methylated sites, the *composite_element_ref*, *gene_symbol*, *entrez_gene_id*, and other features related to the methylated site. For MBV experiments the *beta_value* identifies the information that characterized every single experiment. It is worth noting that redundant fields are repeated for each experiment, because of the sequencing technology and the related chips adopted to generate these data. This allows us to consider two distinct annotations, which describe GEQ and MBV experiments and that we use in order to reduce the size of the data involved in these two types of experiments.

Other experimental data types are available at the OpenGDC public FTP repository, i.e., *Isoform Expression Quantification* [11], *miRNA Expression Quantification* [12], *Masked Somatic Mutation* [13], *Copy Number Segment*, and *Masked Copy Number Segment* [14], but, conversely to GEQ and MBV, every experiment contains different characterizing information. Thus, we could not consider annotations for these cases.

For dealing with data persistency, we adopt a no-SQL document-based Database Management System (DBMS), i.e., MongoDB, to represent all the previously described data types. We use a document-based DBMS both to avoid the problem of being tied to a fixed structure of the data, to vertically and horizontally scale, and to obtain rapid access capabilities. Through this system we are able to represent both semi-structured data as those related to genomic regions, since each type of experiment is defined by specific features, and unstructured data such as clinical data, since the attributes do not follow a well defined schema. Additionally, MongoDB allows us to organize data in different collections, one for each experimental data type. Every collection is defined as a set of documents that are represented as JavaScript Object Notation (JSON) standardized objects. We structured a document to contain one row of the original free-BED files only, which corresponds to an experimental feature on the considered sample. For instance, when taking into account a Copy Number Segment experiment (represented by a specific *.bed* file in the OpenGDC repository) with 355 rows, we added 355 JSON documents to the *Copy Number Segment* collection. In this particular case, every row (i.e., every JSON document) contains a set of information like the genomic coordinates (i.e.: *chromosome*, *start position*, *end position*, and *strand*), the number of probes (*num_probes*), and the segment mean (*segment_mean*), as in the example shown in Listing 1.

It is worth noting that we added three more fields to every document: the experiment identifier *aliquot*, the tumor tag *tumor* (the name of the specific tumor related to the considered experiment), and the *source* field that denotes the data source where data have been extracted. These fields have been added to guarantee the correct reconstruction of the original experimental data.

```
1    {
2        "chrom": "chr1",
3        "start": 62920,
4        "end": 15827002,
5        "strand": "*",
6        "num_probes": 8317,
7        "segment_mean": 0.0031,
8        "aliquot": "01175aae-ce8c-4b95-9293-f73329673009",
9        "tumor": "tcga-acc",
10       "source": "gdc"
11   }
```

Listing 1: JSON representation of a document containing information related to a single row of a Copy Number Segment experiment standardized in free-BED and retrieved from the OpenGDC public FTP repository.

Conversely to all other experimental data types, GEQ and MBV are managed differently. In order to minimize the amount of redundant information we create two additional *annotations* collections. We split every row in the GEQ and MBV free-BED experiments between the *annotation* collection and that one with the information that characterizes the experiment, as described in the above paragraph. We also include clinical and biospecimen information in a separate collection called *metadata*. In this case, every *meta* file in the OpenGDC repository is represented as a document in the *metadata* collection. Here, the procedure of representing these kind of data as documents is pretty simple because of their original key-value structure.

3   Result and Discussion

The proposed data representation allows us to reduce the size of the *bed/tcga* branch of the OpenGDC public FTP repository from ~1.3 TB to ~0.3 TB, producing a gain of ~77% of storage space also thanks to the built-in data compression features provided by the adopted DBMS.

Moreover, to guarantee an easy, fast, and programmatic access to all the information stored and organized in MongoDB, we designed and developed the framework OpenOmics in order to provide a flexible collection of Application Programming Interfaces (APIs). The complete set of implemented endpoints is available at http://bioinformatics.iasi.cnr.it/api/routes. Our APIs are differentiated in three main groups, reflecting the same collections organization of MongoDB. We indeed release (i) a set of endpoints responsible for the interaction with the collections related to the experimental data types, (ii) another group of endpoints able to operate on the annotations collections, and finally (iii) one additional set of endpoints for the interaction with the *metadata* collection. In the following, we describe the main implemented endpoints for each of these groups.

Experiment Endpoints
This set of endpoints is able to interact with the collection containing the experimental data. In particular, it allows to:

i. retrieve the complete list of aliquot ids that represent the whole set of processed experiments:
   `/experiment/source/<source>/program/<program>/tumor/<tumor>/datatype/`
   `<datatype>/aliquots`
ii. extract (a) one single row of the original processed free-BED files or (b) the complete experiment according to the specified data *source*, *program*, *tumor*, *datatype*, *aliquot*, and a particular entity id (e.g.: a methylated site id, a specific ensembl gene id, etc.):
   (a) `/experiment/source/<source>/program/<program>/tumor/<tumor>/`
       `datatype/<datatype>/aliquot/<aliquot>/id/<elem_id>`
   (b) `/experiment/source/<source>/program/<program>/tumor/<tumor>/`
       `datatype/<datatype>/aliquot/<aliquot>/all`

iii. extract a list of overlapping genomics coordinates in a specific experiment according to the specified *chromosome, start position, end position,* and *strand*:

```
/experiment/source/<source>/program/<program>/tumor/<tumor>/datatype/
<datatype>/aliquot/<aliquot>/overlap/chrom/<chrom>/start/<start>/end/
<end>/strand/<strand>
```

Annotation Endpoints

This group of endpoints is related to the GEQ and MBV experimental data only. It allows to interact with the redundant information identified in Sect. 2. In particular, according to a specified *annotation_name* (i.e.: *geneexpression* or *humanmethylation*), it allows to:

i. extract one single document representing the annotation of a specific entity id (i.e.: the methylated site id and the ensembl gene id for the MBV and GEQ data respectively):

```
/annotation/<annotation_name>/id/<elem_id>
```

ii. retrieve the complete annotation:

```
/annotation/<annotation_name>/all
```

Metadata Endpoints

By exploiting this set of endpoints, we are able to extract clinical and biospecimen information related to the experimental data stored in set of *experiments* and *annotations* collections. In particular, these endpoint are able to:

i. retrieve the list of all possible values associate to a specific attribute:

```
/metadata/attribute/<attribute>/all
```

ii. extract the list of all aliquots related to a specific attribute-value couple:

```
/metadata/attribute/<attribute>/value/<value>/aliquots
```

iii. extract a list of all metadata related to a specific *attribute-value* couple:

```
/metadata/attribute/<attribute>/value/<value>/list
```

By releasing these APIs, we aim to extend the features that characterize similar software tools like IRIS-TCGA [15] and [16]. This can represent a preliminary step towards analyses of integrated genomic data as described in [17–20].

4   Conclusions

In this work, we presented an efficient data organization method applied on genomic and clinical data extracted from the OpenGDC FTP repository exploiting the most recent no-SQL technologies. This method allowed us to reduce substantially the size of all the considered data by more than 70%. Additionally, we presented a set of open-source APIs able to facilitate their access and extract potentially significant insights from them. We plan to maintain our system constantly synchronized with the OpenGDC repository, and to extend our database and APIs by modeling new data from other sources. This is straightforward because of the document oriented and non structured format of our data model.

References

1. Stenson, P.D., et al.: The human gene mutation database: towards a comprehensive repository of inherited mutation data for medical research, genetic diagnosis and next-generation sequencing studies. Hum. Genet. **136**(6), 665–677 (2017)
2. Barrett, T., et al.: NCBI GEO: archive for high-throughput functional genomic data. Nucleic Acids Res. **37**(Suppl. 1), D885–D890 (2008)
3. Benson, D.A., Clark, K., Karsch-Mizrachi, I., Lipman, D.J., Ostell, J., Sayers, E.W.: GenBank. Nucleic Acids Res. **42**(D1), D32–D37 (2013)
4. Chen, Q., Zobel, J., and Verspoor, K.: Duplicates, redundancies and inconsistencies in the primary nucleotide databases: a descriptive study. In: Database 2017, baw163 (2017)
5. Cumbo, F., Fiscon, G., Ceri, S., Masseroli, M., Weitschek, E.: TCGA2BED: extracting, extending, integrating, and querying The Cancer Genome Atlas. BMC Bioinform. **18**(1), 6 (2017)
6. Cappelli, E., Cumbo, F., Bernasconi, A., Masseroli, M., Weitschek, E.: OpenGDC: standardizing, extending, and integrating genomics data of cancer. In ESCS 2018: 8th European Student Council Symposium, International Society for Computational Biology (ISCB), p. 1 (2018)
7. Weinstein, J.N., et al.: The cancer genome atlas pan-cancer analysis project. Nat. Genet. **45**(10), 1113 (2013)
8. Jensen, M.A., Ferretti, V., Grossman, R.L., Staudt, L.M.: The NCI genomic data commons as an engine for precision medicine. Blood **130**(4), 453–459 (2017)
9. Mortazavi, A., Williams, B.A., McCue, K., Schaeffer, L., Wold, B.: Mapping and quantifying mammalian transcriptomes by RNA-Seq. Nat. Methods **5**(7), 621 (2008)
10. Bibikova, M., et al.: High density DNA methylation array with single CpG site resolution. Genomics **98**(4), 288–295 (2011)
11. Trapnell, C., et al.: Transcript assembly and quantification by RNA-Seq reveals unannotated transcripts and isoform switching during cell differentiation. Nat. Biotechnol. **28**(5), 511 (2010)
12. Zeng, Y., Cullen, B.R.: Sequence requirements for micro RNA processing and function in human cells. RNA **9**(1), 112–123 (2003)
13. Timmermann, B., et al.: Somatic mutation profiles of MSI and MSS colorectal cancer identified by whole exome next generation sequencing and bioinformatics analysis. PLoS ONE **5**(12), e15661 (2010)
14. Conrad, D.F., et al.: Origins and functional impact of copy number variation in the human genome. Nature **464**(7289), 704 (2010)
15. Cumbo, F., Weitschek, E., Bertolazzi, P., Felici, G.: IRIS-TCGA: an information retrieval and integration system for genomic data of cancer. In: Bracciali, A., Caravagna, G., Gilbert, D., Tagliaferri, R. (eds.) CIBB 2016. LNCS, vol. 10477, pp. 160–171. Springer, Cham (2017). https://doi.org/10.1007/978-3-319-67834-4_13
16. Cumbo, F., Felici, G.: GDCWebApp: filtering, extracting, and converting genomic and clinical data from the Genomic Data Commons portal. In: Genome Informatics, Cold Spring Harbor Meeting (2017)
17. Weitschek, E., Cumbo, F., Cappelli, E., Felici, G.: Genomic data integration: a case study on next generation sequencing of cancer. In: International Workshop on Database and Expert Systems Applications, pp. 49–53, IEEE Computer Society, Los Alamitos (2016)

18. Weitschek, E., Cumbo, F., Cappelli, E., Felici, G., Bertolazzi, P.: Classifying big DNA methylation data: a gene-oriented approach. In: Elloumi, M., et al. (eds.) DEXA 2018. CCIS, vol. 903, pp. 138–149. Springer, Cham (2018). https://doi.org/10.1007/978-3-319-99133-7_11
19. Cappelli, E., Felici, G., Weitschek, E.: Combining DNA methylation and RNA sequencing data of cancer for supervised knowledge extraction. BioData Min. **11**(1), 22 (2018)
20. Weitschek, E., Di Lauro, S., Cappelli, E., Bertolazzi, P., Felici, G.: CamurWeb: a classification software and a large knowledge base for gene expression data of cancer. BMC Bioinform. **19**(10), 245 (2018)

Classification of Pre-cursor microRNAs from Different Species Using a New Set of Features

Malik Yousef[1](✉) and Jens Allmer[2](✉)

[1] Community Information Systems, Zefat Academic College, Zefat 13206, Israel
malik.yousef@gmail.com
[2] Medical Informatics and Bioinformatics,
Hochschule Ruhr West University of Applied Sciences,
Mülheim an der Ruhr, Germany
jens@allmer.de

Abstract. MicroRNAs (miRNAs) are short RNA sequences actively involved in post-transcriptional gene regulation. Such miRNAs have been discovered in most eukaryotic organisms. They also seem to exist in viruses and perhaps in microbial pathogens to target the host. Drosha is the enzyme which first cleaves the pre-miRNA from the nascent pri-miRNA. Previously, we showed that it is possible to distinguish between pre-miRNAs of different species depending on their evolutionary distance using just k-mers.

In this study, we introduce three new sets of features which are extracted from the precursor sequence and summarize the distance between k-mers. These new set of features, named inter k-mer distance, k-mer location distance and k-mer first-last distance, were compared to k-mer and all other published features describing a pre-miRNA. Classification at well above 80% (depending on the evolutionary distance) is possible with a combination of distance and regular k-mer features.

The novel features specifically aid classification at closer evolutionary distances when compared to k-mers only. K-mer and k-mer distance features together lead to accurate classification for larger evolutionary distances such as *Homo sapiens* versus Brassicaceae (93% ACC). Including the novel distance features further increases the average accuracy since they are more effective for lower evolutionary distances. Secondary structure-based features were not effective in this study. We hope that this will fuel further analysis of miRNA evolution. Additionally, our approach provides another line of evidence when predicting pre-miRNAs and can be used to ensure that miRNAs detected in NGS samples are indeed not contaminations. In the future, we aim for automatic categorization of unknown hairpins into all species/clades available in miRBase.

Keywords: microRNA sequence · Machine learning ·
Differentiate miRNAs among species · k-mer miRNA categorization

© Springer Nature Switzerland AG 2019
G. Anderst-Kotsis et al. (Eds.): DEXA 2019 Workshops, CCIS 1062, pp. 15–20, 2019.
https://doi.org/10.1007/978-3-030-27684-3_3

1   Background

Dysregulation of gene expression often defines a disease and microRNAs (miRNAs) are post-transcriptional regulators strongly influencing protein abundance. Mature miRNAs (18–24 nt in length; single stranded) are produced from precursor miRNAs (pre-miRNAs) which are excised from nascent RNA (Erson-Bensan 2014). While miRNAs are described for large parts of the phylogenetic tree, the molecular pathway for plants and animals may have evolved independently (Chapman and Carrington 2007). Both, however, share that pri-miRNAs are transcribed, hairpins (pre-miRNAs) are cleaved from them and that the mature miRNA is incorporated into a protein complex which performs the targeting with the mature sequence as a key element. MicroRNAs have been described for a variety of species ranging from viruses (Grey 2015) to plants (Yousef et al. 2016). Due to involved experimental detection mechanisms, there is reliance on computational approaches to detect miRNAs and many approaches have been developed (Yousef et al. 2006; Allmer and Yousef 2012; Saçar Demirci and Allmer 2014). Many such approaches are based in machine learning and these, with few exceptions (Dang et al. 2008; Yousef et al. 2008; Khalifa et al. 2016), perform two class classification. MicroRNAs and microRNA targets are collected in databases like miRTarBase (Hsu et al. 2014), TarBase (Vergoulis et al. 2012), and MirGeneDB (Fromm et al. 2015) which generally depend on miRBase (Kozomara and Griffiths-Jones 2011) which is the main collection of all miRNAs.

Hundreds of features have been proposed (Saçar Demirci and Allmer 2013) for the parameterization of pre-miRNA sequences. Saçar Demiric et al. (2017) and miRNAfe (Yones et al. 2015) implemented almost all of the published features categorized into sequence, structural, thermodynamic, probabilistic based ones or a mixture of these types which can further be normalized by other features like stem length, number of stems, or similar. The tool, izMiR, evaluated the previously published approaches in terms of their selected feature sets (Saçar Demirci et al. 2017).

Short nucleotide sequences (k-mers) have been used early on for the machine learning-based *ab initio* detection of pre-miRNAs (Lai et al. 2003). Additionally, we have recently conducted studies to answer the question whether the pre-miRNA sequence (ignoring the secondary structure) can be differentiated among species and may, therefore, contain a hidden message that could influence recognition via the protein machinery of the miRNA pathway. We further investigated whether there is a consistent difference among species taking into account their evolutionary relationship.

In order to answer these questions, we established random forest machine learning models using two class classification with the positive class being pre-miRNAs from one species/clade and the negative pre-miRNAs from a different species/clade (Yousef et al. 2017a) and found that distantly related species can be distinguished on this basis. In another recent study (Yousef et al. 2017b), we corroborated on this approach and introduced information-theoretic features but found that k-mers were sufficient for this type of analysis. Here, we have established novel features based on k-mers and compare the performance results with other type of features. The new k-mer distance features perform slightly better (on average ∼1%) than k-mer features and is slightly less effective (on average ∼0.6%) when compared to selected features from all categories. Combining the k-mer distance with the simple k-mer feature sets does not improve performance. However, the novel features are more successful for closer evolutionary distances. In conclusion, the usage of the novel k-mer distance feature set can be encouraged in future studies aiming to differentiate among species based on their miRNAs. In the future, we aim to further analyze the importance of the location of k-mers within a miRNA and the distance among k-mers in order to find a biological interpretation and we will establish an automated categorization system which will place pre-miRNA candidates into their clade/species of origin.

2   Methods

2.1   Parameterization of Pre-miRNAs

Recently, we have shown that k-mers are sufficient to allow categorization of pre-miRNAs into their species of origin (Yousef et al. 2006). Here we use k-mer features as described in (Yousef et al. 2006). In addition to k-mers we use inter k-mer distance where for each k-mer we find its first occurrence in the sequence and then calculate its distance to each k-mer's terminal occurrence in the sequence including the subject k-mer. The sum of these distances computes the overall score which is further normalized by the length of the sequence. Another novel set of features is the k-mer first-last distance which is the distance between the first occurrence and last occurrence of a k-mer within the pre-miRNA sequence. The distance is normalized to the length of the pre-miRNA sequence. Finally, we introduce the k-mer location distance which concerns the average of k-mer distances between locations ($dl = dl/|loci|$). If the k-mer is not found in the sequences the value will be -1 and if it appears only once the value of its feature will be 0. For comparison we also include known secondary structure based features: (1) Number of Base Pairs, Number of Bulges, (2) Number of Loops, (3) Number of bulges with length i ($i = 1$ to 6), (4) Number of bulges with length greater than 6, (5) Number of loops with length I, $i = 1$ to 6 (odd number capture asymmetric loops), and (6) Number of loops with length greater than 6. A KNIME workflow (Berthold et al. 2008) was created to extract those features using the secondary structures obtained from the mirBase (Griffiths-Jones 2010).

The data consists of information from 15 clades. The sequences of Homo sapiens were taken out of the data of its clade Hominidae. The process of removing homology sequences (keeping just one representative) consisted of combining all clades and Homo sapiens sequences into one dataset and then applying the USEARCH (Edgar 2010) to clean the data by removing similar sequences. The USEARCH tool clustered the sequences by similarity. From each cluster, one representative was chosen to form a new dataset with non-homologous sequences. The new dataset was then broken into clades without similar sequences between each pair of clades. Cleaning the data ensured that the results were accurate. The following clades and species from miRBase were used: Hominidae, Brassicaceae, Hexapoda, Monocotyledons (Liliopsida), Nematoda, Fabaceae, Pisces (Chondricthyes), Virus, Aves, Laurasiatheria, Rodentia, Homo sapiens, Cercopithecidae, Embryophyta, Malvaceae, Platyhelminthes.

Following the study of (Yousef et al. 2017a), we used the random forest (RF) classifier implemented by the platform KNIME (Berthold et al. 2008). The classifier was trained and tested with a split into 80% training and 20% testing data. Negative and positive examples were forced to equal amounts using stratified sampling while performing a 100-fold Monte Carlo cross-validation (MCCV) (Xu and Liang 2001) for model establishment. For each established model, we calculated a number of statistical measures like the Matthews's correlation coefficient (Matthews 1975), sensitivity, specificity, and accuracy for evaluation of model performance. All reported performance measures refer to the average of 100-fold MCCVs.

3 Results and Discussion

We have previously shown that k-mers may be sufficient to allow the categorization of miRNAs into species (Yousef et al. 2017b). For this study, we selected pre-miRNAs of a number of species and/or clades to analyze the ability of three new set of features to aid the categorization of pre-miRNAs into their species/clades (see Methods). The selected data represents a range of clades at various evolutionary distances to ensure comprehensive testing. For each pair of species/clades we trained a classifier. We compare k-mer features and all published features with our new feature sets (Table 1).

Table 1. Summary of the pair-wise classification results. Yellow shades indicate lower accuracy while red shades show higher accuracy.

	k-mer	inter k-mer distance	k-mer and inter k-mer distance top 100	k-mer first-last distance	k-mer location distance	Top 100 combined three distance features	All published features	Secondary structure based features
Viruses	0.85	0.84	0.86	0.86	0.86	0.86	0.86	0.74
Monocotyledons	0.81	0.80	0.83	0.81	0.81	0.81	0.83	0.72
Fabaceae	0.82	0.80	0.83	0.81	0.81	0.81	0.82	0.72
Embryophyta	0.85	0.84	0.87	0.86	0.87	0.86	0.87	0.79
Brassicaceae	0.85	0.83	0.86	0.85	0.85	0.85	0.87	0.80
Malvaceae	0.84	0.83	0.85	0.83	0.84	0.84	0.86	0.79
Platyhelminthes	0.82	0.81	0.83	0.82	0.82	0.82	0.85	0.68
Nematoda	0.84	0.82	0.85	0.83	0.83	0.83	0.85	0.73
Hexapoda	0.82	0.80	0.83	0.81	0.82	0.82	0.84	0.69
Pisces	0.81	0.79	0.82	0.81	0.81	0.81	0.82	0.69
Aves	0.81	0.78	0.81	0.80	0.80	0.80	0.81	0.69
Laurasiatheria	0.85	0.84	0.87	0.86	0.86	0.86	0.88	0.82
Rodentia	0.80	0.77	0.80	0.79	0.79	0.79	0.81	0.70
Hominidae	0.77	0.76	0.78	0.77	0.77	0.77	0.78	0.67
Homo sapiens	0.78	0.77	0.79	0.78	0.78	0.78	0.79	0.69
Cercopithecidae	0.77	0.76	0.78	0.77	0.77	0.77	0.79	0.69
Average	0.82	0.80	0.83	0.82	0.82	0.82	0.83	0.73

In summary, secondary structure based features are not as successful when used for categorizing pre-miRNAs into species/clades. Using all published features is as successful as using the best 100 from k-mer and k-mer distance features selected by information gain. K-mer location distance feature is most successful among the k-mer distance features with similar accuracy as using k-mers alone.

4   Conclusions

Here we conducted experiments using a novel transformation of the k-mer features used to parameterize pre-miRNAs for machine learning. Three k-mer distance features, inter k-mer, k-mer first-last, and k-mer location distance were examined and compared to regular k-mer and most published features for pre-miRNA parameterization.

In general, categorization is better for more distant species/clades. The ability of k-mer features to perform accurate categorization at larger evolutionary distances confirms our previous observation (Yousef et al. 2017a, b). K-mer inter and k-mer location distance perform similar to k-mer alone while their combination followed by selection of the best 100 features using information gain leads to a slight increase in average accuracy of 1%. We selected parameters describing the secondary structure of pre-miRNAs in order to understand their contribution for categorization and found that they are on average about 10% less accurate (Table 1). This finding supports the conservation of structure over the conservation of sequence. In conclusion, k-mer and k-mer distance features together lead to accurate categorization for larger evolutionary distances such as Homo sapiens versus Brassicaceae (93% ACC). Including the novel distance features further increases the average accuracy since they are more effective for lower evolutionary distances; while using secondary structure-based features is not as effective. We hope that this will fuel further analysis of miRNA evolution. Additionally, our approach provides another line of evidence when predicting pre-miRNAs and can be used to ensure that miRNAs detected in NGS samples are indeed not contaminations. In the future, we aim for automatic categorization of unknown hairpins into all species/clades available in miRBase.

References

Allmer, J., Yousef, M.: Computational methods for ab initio detection of microRNAs. Front. Genet. **3**, 209 (2012). https://doi.org/10.3389/fgene.2012.00209

Berthold, M.R., Cebron, N., Dill, F., et al.: KNIME: the Konstanz information miner. In: Preisach, C., Burkhardt, H., Schmidt-Thime, L., Decker, R. (eds.) Data Analysis, Machine Learning and Applications, pp. 319–326. Springer, Heidelberg (2008). https://doi.org/10.1007/978-3-540-78246-9_38

Chapman, E.J., Carrington, J.C.: Specialization and evolution of endogenous small RNA pathways. Nat. Rev. Genet. **8**(11), 884–896 (2007). https://doi.org/10.1038/nrg2179

Dang, H.T., Tho, H.P., Satou, K., Tu, B.H.: Prediction of microRNA hairpins using one-class support vector machines. In: 2nd International Conference on Bioinformatics and Biomedical Engineering, iCBBE 2008, pp. 33–36 (2008)

Edgar, R.C.: Search and clustering orders of magnitude faster than BLAST. Bioinformatics **26**(19), 2460–2461 (2010). https://doi.org/10.1093/bioinformatics/btq461

Erson-Bensan, A.E.: Introduction to MicroRNAs in biological systems. In: Yousef, M., Allmer, J. (eds.) miRNomics: MicroRNA Biology and Computational Analysis, 1st edn. Humana Press, New York, pp. 1–14 (2014)

Fromm, B., Billipp, T., Peck, L.E., et al.: A uniform system for the annotation of vertebrate microRNA genes and the evolution of the human microRNAome. Annu. Rev. Genet. **49**, 213–242 (2015). https://doi.org/10.1146/annurev-genet-120213-092023

Grey, F.: Role of microRNAs in herpesvirus latency and persistence. J. Gen. Virol. **96**(4), 739–751 (2015). https://doi.org/10.1099/vir.0.070862-0

Griffiths-Jones, S.: miRBase: microRNA sequences and annotation. Curr. Protoc. Bioinf. 12.9.1–12.9.10 (2010). Chap. 12. Unit. https://doi.org/10.1002/0471250953.bi1209s29

Hsu, S.-D., Tseng, Y.-T., Shrestha, S., et al.: miRTarBase update 2014: an information resource for experimentally validated miRNA-target interactions. Nucleic Acids Res. **42**(Database issue), D78–D85 (2014). https://doi.org/10.1093/nar/gkt1266

Khalifa, W., Yousef, M., Saçar Demirci, M.D., Allmer, J.: The impact of feature selection on one and two-class classification performance for plant microRNAs. PeerJ **4**, e2135 (2016). https://doi.org/10.7717/peerj.2135

Kozomara, A., Griffiths-Jones, S.: miRBase: integrating microRNA annotation and deep-sequencing data. Nucleic Acids Res. **39**(Database issue), D152–D157 (2011). https://doi.org/10.1093/nar/gkq1027

Lai, E.C., Tomancak, P., Williams, R.W., Rubin, G.M.: Computational identification of Drosophila microRNA genes. Genome Biol. **4**(7), R42 (2003). https://doi.org/10.1186/gb-2003-4-7-r42

Matthews, B.W.: Comparison of the predicted and observed secondary structure of T4 phage lysozyme. BBA - Protein Struct. **405**(2), 442–451 (1975). https://doi.org/10.1016/0005-2795(75)90109-9

Saçar Demirci, M.D., Baumbach, J., Allmer, J.: On the performance of pre-microRNA detection algorithms. Nat. Commun. (2017). https://doi.org/10.1038/s41467-017-00403-z

Saçar Demirci, M.D., Allmer, J.: Data mining for microRNA gene prediction: on the impact of class imbalance and feature number for microrna gene prediction. In: 2013 8th International Symposium on Health Informatics and Bioinformatics, pp. 1–6. IEEE (2013)

Saçar Demirci, M.D., Allmer, J.: Machine learning methods for MicroRNA gene prediction. In: Yousef, M., Allmer, J. (eds.) miRNomics: MicroRNA Biology and Computational Analysis SE - 10, 1st edn., pp. 177–187. Humana Press, New York (2014)

Vergoulis, T., Vlachos, I.S., Alexiou, P., et al.: TarBase 6.0: capturing the exponential growth of miRNA targets with experimental support. Nucleic Acids Res. **40**(Database issue), D222–D229 (2012). https://doi.org/10.1093/nar/gkr1161

Xu, Q.-S., Liang, Y.-Z.: Monte Carlo cross validation. Chemom. Intell. Lab. Syst. **56**(1), 1–11 (2001). https://doi.org/10.1016/S0169-7439(00)00122-2

Yones, C.A., Stegmayer, G., Kamenetzky, L., Milone, D.H.: miRNAfe: a comprehensive tool for feature extraction in microRNA prediction. Biosystems **138**, 1–5 (2015). https://doi.org/10.1016/j.biosystems.2015.10.003

Yousef, M., Allmer, J., Khalifa, W.: Plant microRNA prediction employing sequence motifs achieves high accuracy (2016)

Yousef, M., Jung, S., Showe, L.C., Showe, M.K.: Learning from positive examples when the negative class is undetermined–microRNA gene identification. Algorithms Mol. Biol. **3**, 2 (2008). https://doi.org/10.1186/1748-7188-3-2

Yousef, M., Khalifa, W., Acar, I.E., Allmer, J.: MicroRNA categorization using sequence motifs and k-mers. BMC Bioinf. **18**(1), 170 (2017a). https://doi.org/10.1186/s12859-017-1584-1

Yousef, M., Nebozhyn, M., Shatkay, H., et al.: Combining multi-species genomic data for microRNA identification using a Naive Bayes classifier. Bioinformatics **22**(11), 1325–1334 (2006). https://doi.org/10.1093/bioinformatics/btl094

Yousef, M., Nigatu, D., Levy, D., et al.: Categorization of species based on their MicroRNAs employing sequence motifs, information-theoretic sequence feature extraction, and k-mers. EURASIP J. Adv. Sig. Process. **2017**(70), 1–10 (2017b). https://doi.org/10.1186/s13634-017-0506-8

A Review of Quasi-perfect Secondary Structure Prediction Servers

Mirto Musci[(✉)][iD], Gioele Maruccia[iD], and Marco Ferretti[iD]

Universita' di Pavia, 27100 Pavia, Italy
mirto.musci@unipv.it

Abstract. The secondary structure was first described by Pauling et al. in 1951 [14] in their findings of helical and sheet hydrogen bounding patterns in a protein backbone. Further refinements have been made since then, such as the description and identification of first 3, then 8 local conformational states [10]. The accuracy of 3-state secondary structure prediction has risen during last 3 decades and now we are approaching to the theoretical limit of 88–90%. These improvements came from increasingly larger databases of protein sequences and structures for training, the use of template secondary structure information and more powerful deep learning techniques. In this paper we review the best four scorer servers which provide the highest accuracy for 3- and 8-state secondary structure prediction.

Keywords: Secondary structure prediction · Deep neural networks · Machine learning

1 Introduction

Artificial intelligence is playing an increasingly important role in the development of Bioinformatics [12]. Many improvements have been made in fields such as gene therapy, personalized medicine, preventative medicine, drug development and biotechnology [22]. In order to apply many of these applications, the native structure of the protein must be known [21].

1.1 Background

In early 1970s, Anfisen demonstrated that the native tertiary structure is encoded in the primary structure [2]. From then on, several methods have been developed to predict the structure from the sequence. Unfortunately, this is not an easy task without using known structures as templates [24,28]. That is why the protein structure prediction problem has been decomposed into smaller sub-problems to ease and improve the entire resolution process.

The most well-known sub-problem is the prediction of the protein secondary structure, that is assigning to the amino-acid sequence a list of secondary structure types. Indeed, the accuracy of protein secondary structure prediction

© Springer Nature Switzerland AG 2019
G. Anderst-Kotsis et al. (Eds.): DEXA 2019 Workshops, CCIS 1062, pp. 21–26, 2019.
https://doi.org/10.1007/978-3-030-27684-3_4

directly impacts the accuracy of tertiary one. Secondary structure prediction must be done in either 3-state classification or 8-state classification (Sect. 1.4). Despite more that an half century of history, the accuracy of this process continues to grow actively. Since then, secondary structure prediction has been reviewed periodically [13,15,17,26].

The intent of this paper is to provide a non-exhaustive update to [26] in order to show the most recent methods for secondary structure prediction which are approaching, at least for the 3-state classification problem, to the theoretical accuracy limit. This limit, imposed on secondary structure prediction, is the somewhat arbitrary definition of the three states considered: ideal helices and sheets do not exist and there are no clear boundaries between helix and coil nor sheet and coil states. It was shown that structural homologies differ by about 12% in secondary structure assignment [19] for those with >30% sequence identity [27].

1.2   Dataset

This paper compares all the reviewed approaches on the same dataset, the one employed by Hanson et al. for the accuracy comparison with SPOT-1D [7].

To facilitate a fair comparison to other methods, it employed structures coming from PDB [3] in the first 7 months of 2018, redundancy filtered, constrained at <2.5A where an identity cutoff of 25% has been made in order to minimize evaluation bias of overlapping data.

1.3   Multiple Sequence Alignment

Multiple sequence alignments were introduced to the prediction of secondary structure in early 1990s and is a critical preprocessing step for any modern prediction algorithm.

Using multiple sequence alignment information, rather than only protein sequence, has led to a 10% accuracy improvement [18]. The strength of including multiple sequence alignment information in prediction is the evolutionary information they contain, which is much richer than a single sequence. The main evolutionary information algorithms which aim to this purpose are PSI-BLAST (Position-Specific Iterated BLAST)[1] and HHblits [16]. They work on the principle of position specific scoring matrix (PSSM) and profile Hidden Markov Models (HMMs), respectively.

1.4   SS3 and SS8 Accuracy

Protein secondary structure prediction is often evaluated by the SS3 and SS8 accuracy: the first is a three-class classification, i.e., helix, strand and coil, while the second, more specific and difficult to increase, is the accuracy of an eight-class classification: 310-helix, $\alpha - helix$, $\pi - helix$, $\beta - strand$, $\beta - bridge$, $\beta - turn$, bend, and loop or irregular.

Only accuracy values will be provided for the comparison of the following 4 methods. However, further specifications and metrics can be found in the literature.

2 Methods

As of today, the best performances are obtained from research groups who are making prediction servers available online. We will describe the best four of them: SPOT-1D (Subsect. 2.1), NetSurfP-2.0 (Subsect. 2.2), MUFOLD (Subsect. 2.3) and Porter5 (Subsect. 2.4), listed in descendent order. We summarize their architecture, provide location of their implementation, and briefly discuss their predictive performances.

They adopt sophisticated deep learning methods to represent the relationship between the amino-acid input sequence (or, more precisely, the evolutionary information encoded in its PSSM) and the secondary structure states. Prediction methods are provided to the end users as standalone applications and/or as web servers, depending by the amount of data that is going to be predicted and the computational power needed.

2.1 SPOT-1D

Architecture: Ensemble of LSTM-BRNN, ResNets, Fully-Connected NN
Availability: http://sparks-lab.org/index.php/Main/Services

SPOT-1D, developed by Hanson et al. [7] in his last Raptor update, adopts a deep learning architecture that has the aim to counteract lacks of previous methods. Some of them, like NetSurfP-2.0 [11] and MUFOLD-SS [6], adopt a large Long Short-Term Memory (LSTM) network [9] in a Bidirectional Recurrent Neural Network (BRNN) [20] and variants of Inception networks, respectively [23]. Other methods, like Porter5, employ an ensemble of BRNN [25] and, moreover, deep convolutional neural fields (DeepCNF) and deep residual convolutional neural networks (ResNets) [8] have been employed by previous method to fulfill prediction goals.

What Hanson et al. pointed out is that all the above methods for making secondary structure prediction that rely on the separate application of different individual neural networks, can't exploit the different capabilities in capturing local and/or non-local interactions of each network. They demonstrate that, by using an ensemble of models based on LSTM-BRNN, ResNet, and Fully-Connected (FC) NN, allows a significant improvement in the prediction of residue-residue contact map, which will be the input of secondary structure prediction model. Indeed, previous studies have shown the improvement in secondary structure prediction resulting from the input of native contact maps [4,5].

2.2 NetSurfP-2.0

Architecture: CNN, LSTM, FC
Availability: http://www.cbs.dtu.dk/services/NetSurfP-2.0

NetSurfP-2.0 method [11], developed by Klausen et al., adopts a complex network structure that, as other methods do, solves not only the problem of secondary structure prediction but also other related protein structure properties. Indeed, the input layer of the network consists of the encoded sequence plus the full HMM profiles, consisting in a total of 50 input features. This input is connected to 2 convolutional neural network (CNN) layers, which output is concatenated with the initial 50 input features and connected to 2 bidirectional long short-term memory (LSTM) layers. Each output is then calculated with a fully connected (FC) layer using the results of the final LSTM layer.

2.3 MUFOLD

Architecture: Deep Inception-Inside-Inception network
Availability: http://dslsrv8.cs.missouri.edu/~cf797/MUFoldSS/download.html

Developed by Fang et al. [6] with the purpose of creating a method that could extract non-local interactions of residues by maintaining efficient computation times and memory. This architecture makes use of a sequence of inception modules followed by several convolutional and fully connected dense layers.

The inception module, a state-of-the-art image recognition method, consists of multiple convolutional operations in parallel and concatenates the resulting feature maps before going into next layer. By using these inception modules as building blocks for the final deep inception-inside-inception network, which extends deep inception networks through nested inception modules. More clear details are present in the related article.

2.4 Porter5

Architecture: BRNN, CNN
Availability: http://distilldeep.ucd.ie/porter/

Porter 5 [25], similarly to the approach of SPOT-1D, selects an ensemble of neural networks, even though they come from the same family. It is based on ensemble of 7 Bidirectional Recurrent Neural Networks (BRNN), 3 trained on PSI-BLAST, 3 trained on HHblits and 1 trained on both. The implementation of Torrisi et al., the authors, consists of two similar cascaded stages, both of which contain a classic BRNN layer followed by a convolutional layer.

3 Results

Due to the fair and complete comparison made by Hanson et al. [7], here we just present their results in a compact and focused way in Table 1. As we said, this article doesn't provide an exhaustive comparison between all working methods used these days, but only best ones are compared and discussed.

Table 1. Test performance on best 4 prediction servers on the latest PDB structures. All classification SS3 and SS8 accuracies are presented in % accuracy.

Predictor	SS3 accuracy	SS8 accuracy
SPOT-1D	86.18	75.41
NetSurfP-2.0	85.31	73.81
MUFOLD	84.78	73.66
Porter5	84.78	73.22

References

1. Altschul, S., et al.: Gapped blast and PSI-blast: a new generation of protein databases search programs. Nucleic Acids Res. **25**, 3389–402 (1997). https://doi.org/10.1093/nar/25.17.3389
2. Anfinsen, C.B.: Principles that govern the folding of protein chains. Science **181**(4096), 223–230 (1973)
3. Bourne, P.E.: The protein data bank. In: Protein Structure: Determination, Analysis, and Applications for Drug Discovery, p. 389 (2003)
4. Ceroni, A., Frasconi, P.: On the role of long-range dependencies in learning protein secondary structure. In: 2004 IEEE International Joint Conference on Neural Networks (IEEE Cat. No. 04CH37541), vol. 3, pp. 1899–1904. IEEE (2004)
5. Ceroni, A., Frasconi, P., Pollastri, G.: Learning protein secondary structure from sequential and relational data. Neural Netw. **18**(8), 1029–1039 (2005)
6. Fang, C., Shang, Y., Xu, D.: MUFOLD-SS: new deep inception-inside-inception networks for protein secondary structure prediction. Proteins: Struct. Funct. Bioinform. **86**(5), 592–598 (2018)
7. Hanson, J., Paliwal, K., Litfin, T., Yang, Y., Zhou, Y.: Improving prediction of protein secondary structure, backbone angles, solvent accessibility and contact numbers by using predicted contact maps and an ensemble of recurrent and residual convolutional neural networks. Bioinformatics (2018)
8. He, K., Zhang, X., Ren, S., Sun, J.: Deep residual learning for image recognition. In: Proceedings of the IEEE Conference on Computer Vision and Pattern Recognition, pp. 770–778 (2016)
9. Hochreiter, S., Schmidhuber, J.: Long short-term memory. Neural Comput. **9**(8), 1735–1780 (1997)
10. Kabsch, W., Sander, C.: Dictionary of protein secondary structure: pattern recognition of hydrogen-bonded and geometrical features. Biopolymers **22**(12), 2577–2637 (1983)
11. Klausen, M.S., et al.: NetSurfP-2.0: improved prediction of protein structural features by integrated deep learning. Proteins Struct. Funct. Bioinform. **87**, 520–527 (2019)
12. Moult, J., Fidelis, K., Kryshtafovych, A., Schwede, T., Tramontano, A.: Critical assessment of methods of protein structure prediction (CASP) - round XII. Proteins: Struct. Funct. Bioinform. 86(Suppl. 1) (2017). https://doi.org/10.1002/prot.25415
13. Oldfield, C.J., Chen, K., Kurgan, L.: Computational prediction of secondary and supersecondary structures from protein sequences. In: Kister, A. (ed.) Protein Supersecondary Structures, pp. 73–100. Springer, New York (2019). https://doi.org/10.1007/978-1-4939-9161-7_4

14. Pauling, L., Corey, R.B., Branson, H.R.: The structure of proteins: two hydrogen-bonded helical configurations of the polypeptide chain. Proc. Natl. Acad. Sci. **37**(4), 205–211 (1951). https://doi.org/10.1073/pnas.37.4.205. https://www.pnas.org/content/37/4/205

15. Pirovano, W., Heringa, J.: Protein secondary structure prediction. In: Carugo, O., Eisenhaber, F. (eds.) Data Mining Techniques for the Life Sciences, pp. 327–348. Springer, Heidelberg (2010). https://doi.org/10.1007/978-1-60327-241-4_19

16. Remmert, M., Biegert, A., Hauser, A., Söding, J.: HHblits: lightning-fast iterative protein sequence searching by HMM-HMM alignment. Nat. Methods **9**(2), 173 (2012)

17. Rost, B.: Protein secondary structure prediction continues to rise. J. Struct. Biol. **134**(2–3), 204–218 (2001)

18. Rost, B., Sander, C.: Improved prediction of protein secondary structure by use of sequence profiles and neural networks. Proc. Nat. Acad. Sci. **90**(16), 7558–7562 (1993)

19. Rost, B., Sander, C., Schneider, R.: Redefining the goal of protein secondary structure prediction. J. Mol. Biol. **235**, 13–26 (1994). https://doi.org/10.1016/S0022-2836(05)80007-5

20. Schuster, M., Paliwal, K.K.: Bidirectional recurrent neural networks. IEEE Trans. Sig. Process. **45**(11), 2673–2681 (1997)

21. Staples, M., Chan, L., Si, D., Johnson, K., Whyte, C., Cao, R.: Artificial intelligence for bioinformatics: applications in protein folding prediction (2019). https://doi.org/10.1101/561027

22. Stephenson, N., et al.: Survey of machine learning techniques in drug discovery. Curr. Drug Metab. **19** (2018). https://doi.org/10.2174/1389200021966618082011 2457

23. Szegedy, C., Ioffe, S., Vanhoucke, V., Alemi, A.A.: Inception-v4, inception-resnet and the impact of residual connections on learning. In: Thirty-First AAAI Conference on Artificial Intelligence (2017)

24. Tai, C.H., Bai, H., Taylor, T.J., Lee, B.: Assessment of template-free modeling in CASP10 and ROLL. Proteins Struct. Funct. Bioinform. **82**, 57–83 (2014)

25. Torrisi, M., Kaleel, M., Pollastri, G.: Porter 5: fast, state-of-the-art AB initio prediction of protein secondary structure in 3 and 8 classes. bioRxiv, p. 289033 (2018)

26. Yang, Y., et al.: Sixty-five years of the long march in protein secondary structure prediction: the final stretch? Brief. Bioinform. **19**(3), 482–494 (2016)

27. Zhang, W., Dunker, A.K., Zhou, Y.: Assessing secondary structure assignment of protein structures by using pairwise sequence-alignment benchmarks. Proteins: Struct. Funct. Bioinform. **71**(1), 61–67 (2008). https://doi.org/10.1002/prot.21654. https://onlinelibrary.wiley.com/doi/abs/10.1002/prot.21654

28. Zhou, Y., Duan, Y., Yang, Y., Faraggi, E., Lei, H.: Trends in template/fragment-free protein structure prediction. Theor. Chem. Acc. **128**(1), 3–16 (2011)

Theoretical and Practical Analyses in Metagenomic Sequence Classification

Hend Amraoui[1,2,3], Mourad Elloumi[3], Francesco Marcelloni[1], Faouzi Mhamdi[3], and Davide Verzotto[1,4,5(✉)]

[1] Department of Information Engineering, University of Pisa, Pisa, Italy
[2] University of Tunis El Manar, Tunis, Tunisia
[3] Laboratory of Technologies of Information and Communication and Electrical Engineering (LaTICE), National Higher School of Engineers of Tunis (ENSIT), University of Tunis, Tunis, Tunisia
[4] Institute for Informatics and Telematics, CNR, Pisa, Italy
[5] Euro-Mediterranean Biomedical Scientific Institute (ISBEM), Pisa and Mesagne, Italy
amraoui.hend@yahoo.fr, mourad.elloumi@gmail.com,
francesco.marcelloni@unipi.it, faouzi.mhamdi@ensi.rnu.tn,
davide.verzotto@iit.cnr.it

Abstract. Metagenomics is the study of genomic sequences in a heterogeneous microbial sample taken, e.g. from soil, water, human microbiome and skin. One of the primary objectives of metagenomic studies is to assign a taxonomic identity to each read sequenced from a sample and then to estimate the abundance of the known clades. With ever-increasing metagenomic datasets obtained from high-throughput sequencing technologies readily available nowadays, several fast and accurate methods have been developed that can work with reasonable computing requirements. Here we provide an overview of the state-of-the-art methods for the classification of metagenomic sequences, especially highlighting theoretical factors that seem to correlate well with practical factors, and could therefore be useful in the choice or development of a new method in experimental contexts. In particular, we emphasize that the information derived from the known genomes and eventually used in the learning and classification processes may create several experimental issues—mostly based on the amount of information used in the processes and its uniqueness, significance, and redundancy,—and some of these issues are intrinsic both in current alignment-based approaches and in compositional ones. This entails the need to develop efficient alignment-free methods that overcome such problems by combining the learning and classification processes in a single framework.

Keywords: Metagenomic sequence classification ·
Alignment-free algorithms · Genome analysis · Combinatorics ·
Pattern discovery · Strings

H.A. is supported and M.E. and D.V. are partially supported by the ERASMUS+ KA107 project no. 2018-1-IT02-KA107-047786.

1   Introduction

A major challenge in the analysis of microbial samples is the identification of all clades present at different taxonomic levels and their relative abundances, in order to understand microbial community effects and recognize potential pathogens in the analysis of soil, water, and other environments [17], and the study of the human health, from human microbiome to host-microbial co-metabolic interactions [12].

In modern metagenomic approaches, next-generation sequencing technologies (NGS) are applied to a sample and the generated reads are classified to estimate the presence and abundance of all known clades, from domain to strain levels. Two characteristic phases are typically identified to solve the problem: (i) learning from known, large reference genome sequences and (ii) comparing and classifying every single read. Both phases might include (sub)sequence analysis and extraction either from, possibly thousands, known genomes, e.g. of size in the order of 10^6 base pairs (bp) for bacterial genomes (or of 10^4 bp for the genetic part only), or from each read, of size of 10^2 bp and typically of $10^7 - 10^8$ in number.

A number of methods have been proposed for metagenomic sequence classification that trade-off computational complexity for predictive power [14] in the analysis of NGS data. These methods address some of the big data issues, such as the large volume of sequenced reads, sequencing errors, and the incompleteness of the reference genome databases. In addition, there might be classification issues for highly similar clades at lower taxonomic levels, as well as for divergent clades otherwise sharing similar genome sequences [9].

We here analyze several the state-of-the-art methods for metagenomic sequence classification using NGS data from both the theoretical and the experimental points of view, in order to highlight strengths and weaknesses of each type of methods. The objective is to delineate some quantitative properties of the methods and correlate them with experimental results, to help solving important information content issues and improve the complexity–prediction trade-off with more suitable approaches in the future.

Three types of strategies are usually adopted to solve the metagenomic sequence classification problem:

- *alignment-based methods* (a), which perform a complete analysis by fully aligning all reads to the known genomes;
- *marker-based methods* (m), which use smaller databases of marker sequences that specifically identify the clades;
- *k-mer-based methods* (k), which analyze and extract fixed-size patterns, called *k*-mers, either in the genomes or in the read.

K-mer-based approaches typically follow alignment-free strategies, which go beyond the concept of sequence alignment and are thus becoming increasingly popular in the analysis of massive data in bioinformatics [5,6,8], although more efficient alignment-based approaches have been also recently proposed [3,19,22,23]. We refer the reader to [4,7,25] and for a more complete review on alignment-free and pattern analysis algorithms on strings.

In the following, we will first analyze the information content used by the methods from a theoretical point of view, and then correlate it with practical factors in the experimental results.

2 Information Content Analysis

Let us consider strings with characters (or bases) from a finite alphabet Σ of size σ. A substring is a pattern defined as a contiguous subsequence of characters within a string. A k-mer is a substring of fixed length k, while a spaced k-mer is a subsequence with a fixed number k of characters intermixed by a fixed number e of possible mismatches (or wild cards). An extensible pattern is a subsequence with a variable number of possible mismatches per position, similar to sequence alignment results, which can be quantified by its minimum length with errors.

We would like to map a string r of length l, the read, to a set of m strings g_i, with $1 \leq i \leq m$, representing the known genome sequences, for simplicity considered all of length n. A common pattern is any subsequence that is in common to both r and any of the g_i. The approaches here studied analyze sets of common patterns to decide for the best read classification or call for ambiguity or non classified read.

To analyze the information content we considered the case of perfect information, in which the read r must belong exactly to one and only one of the known genomes; i.e. the read must be both a substring of it and unambiguously classifiable. This concept could be further extended to the analysis of clades at any taxonimic level, by implicitly considering the representative genomes of the taxa in each clade and the average length n of them. As a side note, considering the case of non-perfect information would require a unique definition of fuzzy identity between the read and the genomes, which is not readily available (e.g. by considering genomic rearrangements). Beside, most methods perform optimizations based on different measures of identity and this would make the comparison fair.

We defined four information content indicators and analyzed their values by considering the theoretical worst case for each method. For example, we typically considered the case in which all known genomes are identical to each other, except in $\lceil \log_\sigma m \rceil \ll n$ bases, as depicted in Fig. 1a. The four indicators are as follows:

- *encoded information:* the total amount of unique bases analyzed in the genomes and used both in the learning and classification phases;
- *pattern matching power:* the maximum length that could be attained by a common pattern identified between a genome and the read, which can be considered as a function of statistical significance in the analysis of stochastic processes;
- *redundancy:* the total number of single-base overlaps between common patterns of a specific genome and the read (see [4] for a more formal definition);
- *non-uniqueness:* the total length of non-unique common patterns from all genomes that are accounted for in the classification phase.

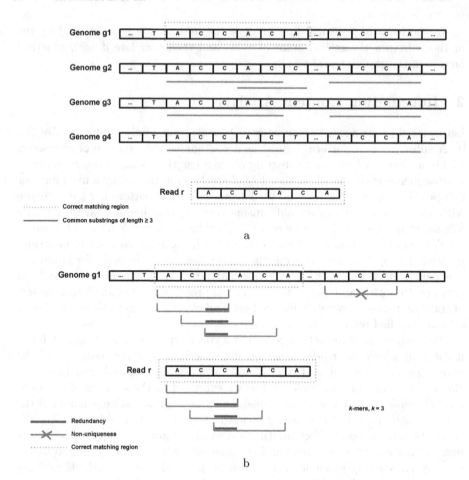

Fig. 1. Metagenomic sequence classification: (a) a worst-case scenario in case of perfect information, with four known genomes that are identical to each other, except in one base (in light blue); (b) a conceptual overview of the inclusion of redundant and non-unique common patterns in the learning and classification phases for some k-mer-based approaches (e.g. Kraken). (Color figure online)

This is the first study to systematically quantify all such theoretical effects caused by several trade-offs employed by the methods, and to subsequently correlate them with practical factors in experimental contexts. These indicators are analyzed asymptotically, and also not normalized to the encoded information to better reflect the peculiarities of the individual strategies. Quality factors in each method, instead, will not be explicitly taken into account, because they largely depend on specific notions and they would be typically hard to be actually quantified.

Table 1. Worst-case theoretical information content: state-of-the-art methods sorted first by type, and then by higher encoded information and pattern matching power, and by lower redundancy and non-uniqueness.

Method	Type	Encoded information	Pattern matching power	Redundancy	Non-uniqueness
BLAST-MEGAN	a	$O(nm)$	l	none	none
DIAMOND-MEGAN	a	$O(nm)$	l	none	none
MetaPhlAn2	m (a)	$O(r_g nm)$	l	none	none
GOTTCHA	m (k)	$O(r_g nm)$	k	none	$O(r_g knm)$
LMAT	k	$O(nm)$	k	$O(kn)$	$O(lkm)$
MetaVW	k	$O(nm)$	k	$O(kn)$	$O(lkn)$
Kraken	k	$O(nm)$	k	$O(kn)$	$O(lknm)$
MSC	k (a)	$O((\log_\sigma m)m)$	$\max\{k, \log_\sigma m\}$	none	none
CLARK-S	k	$O((\log_\sigma m)m)$	$k+e$	$O(k^{e+1}\sigma^e n)$	$O(lk^{e+1}\sigma^e n)$
CLARK	k	$O((\log_\sigma m)m)$	k	$O(kn)$	$O(lkn)$
KrakenUniq	k	$O((\log_\sigma m)m)$	k	$O(kn)$	$O(lkn)$
SKraken	k	$O((\log_\sigma m)m)$	k	$O(kn)$	$O(lkqnm)$
CLARK-l	k	$O(r_s km)$	k	none	$O(kn)$

Legend: l: size of a read, m: number of clades, n: average length of representative genome sequences in a clade, r_g: ratio of marker sequences in the genomes (e.g. 0.01), r_s: sampling rate for non-overlapping patterns, k: k-mer length, σ: size of the alphabet (e.g. 4 for DNA), e: error rate allowed in spaced k-mers, q: non-uniqueness threshold.

Indeed, we would like to analyze how positive indicators such as the encoded information and the pattern matching power, both requiring more computation resources, and negative indicators such as redundancy and non-uniqueness, affect each strategy and therefore reflect into experimental analysis. In fact, if the learning process has to deal with a set of redundant or non-unique common patterns, this could increase false positive discoveries, because there might be available no immediate controls to the effects of these patterns in the decision process, especially in real context with non-perfect information. On the other hand, when both the learning and the classification phases can deal with the maximum information available, $O(nm)$, and the highest pattern matching power, l, this could especially increase the sensitivity and resolve the classification process even for sequences hardly to map, largely helping the annotation at lower taxonomic levels. Nevertheless, the quantitative indicators here proposed could be all correlated with the classical predictive and computational complexity measures, as an effect of the trade-offs proposed by every method.

In Table 1, we analyze the theoretical information content of a set of 13 metagenomic sequence classification tools in the state-of-the-art, based on the different types of methods (a, m, k, with some mixed versions thereof), by using the four indicators and the respective worst cases, as specified in the following.

Alignment-Based Methods. Alignment-based methods are the reference methods for accurately aligning all reads to the known genomes, by performing a complete metagenomic analysis. Examples of these are BLAST-MEGAN [11] and DIAMOND-MEGAN [3], for which we can consider the obtained alignments as extensible common patterns. They indeed deal with the maximum encoded information, $O(nm)$, by including whole genomes in the learning phase, and account for the highest pattern matching power, l, by aligning fully each read to the genomes. This conveys no redundancy and no non-uniqueness in the information used in the process.

Marker-Based Methods. Marker-based-methods select only a small fraction of the genome, r_g (e.g. 1%), based either on the genetic material only or on a mix of the genetic and the non-genetic material that is well conserved within each clade and unique between different clades, and can further follow either alignment-based (e.g. MetaPhlAn2 [20]) or k-mer-based (e.g. GOTTCHA [9]) strategies.

MetaPhlAn2 is based on unique marker genes and provides the highest possible pattern matching power l, with no redundancy and no non-uniqueness accounted for, similar to the classical alignment-based methods.

GOTTCHA, instead, splits the read into non-overlapping k-mers and maps them (with an alignment process) to, possibly overlapping, patterns in unique marker sequences extracted from a mix of genetic and non-genetic materials of the known genomes. This leads to a more flexible process, but limits the pattern matching power to k. Although some highly conservative filters are applied to limit the false discovery rate, in the worst case such that all the genomes and the read are mostly composed by the same symbol $x \in \Sigma$, this introduces a non-uniqueness factor of $O(r_g knm)$ due to overlapping non-unique patterns in the genomes, but keeps the redundancy to none by avoiding the k-mer overlaps in the read.

K-Mer-Based Methods. K-mer-based methods are characterized by the extraction of all the possible k-mers from the whole genomes and the subsequent comparison with those extracted from the read, accounting for a limited pattern matching power of k, similar to GOTTCHA.

Two kinds of approaches are typically available, those that include the entire information from the genomes, $O(nm)$ (e.g. LMAT [1], MetaVW [21], and Kraken [24]), and those that retain only the informative and/or unique k-mers, (e.g. CLARK methods [15,16], MSC [18], KrakenUniq [2], and SKraken [13]) which account, instead, for $O((\log_\sigma m)m)$ information in the worst case, based on some measures.

A redundancy effect of $O(kn)$ is present in all methods that account for overlapping k-mers and their frequencies in both the genomes and the read, as in the cases of LMAT, MetaVW, Kraken, CLARK, KrakenUniq, and SKraken. CLARK-S further permits for both a higher pattern matching power of $k+e$ and for more flexibility, by allowing up to e mismatches in spaced patterns, which could inflate the redundancy of a factor of $k^e \sigma^e$. CLARK-l [16], instead, is a lighter version of CLARK, which reduces the encoded information by sampling the informative patterns by the rate of r_s, while avoiding overlaps between the sampled patterns in the known genomes, indeed eliminating any redundancies.

By considering the worst case of having both the genomes and the read mostly composed by the same symbol $x \in \Sigma$, Kraken seems to largely account

Table 2. (a) Predictive power for taxa detection at different taxonomic levels: precision (P), recall (R), and F1-score (in bold are the best values for each column), and (b) computational resources required (run time and RAM), derived from the results of [14].

Method	Genus			Species			Strain			Method	Run time	Memory (GB)
	P	R	F1	P	R	F1	P	R	F1			
BLAST-MEGAN	94	95	**94**	**99**	94	**92**	70	56	64	BLAST-MEGAN	> 1 week	16
DIAMOND-MEGAN	96	98	92	86	90	**92**	90	16	22	DIAMOND-MEGAN	2 days	24
MetaPhlAn2	94	84	90	90	84	80	**100**	24	39	MetaPhlAn2	30 min	8
GOTTCHA	**99**	88	92	90	88	**92**	64	62	**65**	GOTTCHA	40 min	18
LMAT	41	**100**	58	26	96	44	21	**88**	30	LMAT	1 h	25
Kraken	58	**100**	70	42	98	60	20	76	28	Kraken	40 min	80
CLARK-S	16	98	34	18	**100**	28	-	-	-	CLARK	30 min	87
CLARK	16	96	28	14	**100**	22	-	-	-	CLARK-S	2 h	>100

a   b

for all possible non-unique common patterns, $O(lknm)$, with SKraken decreasing this value by a factor of q (an input threshold for disambiguating k-mers in clades), and MetaVW further reducing it by accounting for k-mer spectrum distances within each clade and across different clades of the known genomes, in the learning phase. KrakenUniq and CLARK, instead, reduce original Kraken non-uniqueness by retaining the k-mers unique to each clade only, but still account for intragenome non-unique k-mers. CLARK-S amplifies the non-uniqueness effect due to intrinsic, intragenome k-mer overlaps of the spaced seeds, while CLARK-l reduces it to $O(kn)$ by eliminating intragenome overlaps, but not the k-mer overlaps in the read. Vice versa, LMAT uses a binary approach in the construction of k-mer-based taxonomic trees (i.e. without k-mer frequencies), to avoid any genome representation biases, indeed reducing the non-uniqueness to $O(lkm)$; however, this could also partly reduce the encoded information when either the read complexity is low or its length is large.

Unlike the other methods of this type, MSC first extracts unique k-mers from the known genomes and then tries to assemble them, possibly increasing the pattern matching power to $log_\sigma m$ for each assembled segment. Performing a subsequent alignment-based approach by mapping the read to the segments, it turns off both redundancy and non-uniqueness effects. In this way, MSC does not carry over non-unique information in the decisional process.

3   Correlation with Experimental Results

To experimentally compare the methods and the main types of strategies, in Table 2a we provide a summary of the median values of precision, recall, and F1-score for the detection of taxa at any abundance at the taxonomic levels of genus, species, and strain, as derived from the experiments in McIntyre, *et al.* [14]. In Table 2b we further provide the worst-case computational resources required by the tools in the benchmark. These in-depth analyzes seem to be generally considered as the most complete and robust done so far, and include both different synthetic scenarios at the extreme ends of complexity and laboratory-generated datasets that typically fit between the synthetic scenarios, but with some peculiarities. Although not all tools are included in this comparison, especially the most recent, most of the theoretical features of these methods previously explained can be mapped to the experimental results.

An overall comparison is made regarding both clade detection and relative abundance estimation in the entire metagenomic sample, without measuring the classification of individual readings. This is mainly due to the fact that methods with low encoded information could classify only a small part of the metagenomic dataset, and, furthermore, we are usually only interested in the overall sample analysis. Moreover, to correct great false positive discoveries when the default parameters are merely relaxed, we considered the filtered versions for some of the approaches, as provided by [14].

According to Table 2a, the reference, alignment-based methods seem to provide the best values for all measures at all taxonomic levels, beside a drop in DIAMOND-MEGAN strain-level recall, probably because of the strict index optimizations used there. However, these methods also involve high execution times, which diminish their usability in everyday applications (see Table 2b). These factors can be easily explained by the theoretically maximum encoded information, $O(nm)$, and the highest pattern matching power, l, with no redundancy nor non-uniqueness involved, as previously expected.

Marker-based methods had the worst recall, especially at the strain level, but some of the highest precision and F1 values, as well as some of the best running times and memory usages. This can be mostly explained by a lower encoded information with respect to most methods (beside CLARK tools) and no redundancy, as well as a conservative analysis on the selected information. On the other hand, the flexibility provided by a lower pattern matching power seems to favor GOTTCHA with respect to MetaPhlAn2 in recall at the strain level, while penalizing it in precision there, when also the non-uniqueness factor likely comes into play. The inclusion of non-genetic material as well and some ad-hoc, conservative post-processing analyses in GOTTCHA yielded the best F1-score at the strain level, although more in-depth analyses might reject its conservativeness too high due to great accuracy drops in the abundance estimation [14].

K-mer-based methods of the first subtype (LMAT and Kraken) typically get the highest levels of recall, mainly comparable to BLAST-MEGAN, because of the maximum encoded information similar to alignment-bases approaches, and of the flexibility given by the lower pattern matching power of k with respect to marker based-methods. On the other hand, CLARK tools seem to largely trade-off recall with precision, and thus apparently result in pretty high recalls, even if they should retain discriminative k-mers only. Notwithstanding, this does not seem to be feasible anymore as long as CLARK tools analyze the lowest taxonomic levels. Moreover, a lower pattern matching power in addition to a high degree of both redundancy and non-uniqueness seem to drastically reduce the precision of k-mer-based methods with respect to all the other types of strategies, yielding the lowest precision and F1 values (e.g. for CLARK tools). The amount of redundancy, and possibly of non-uniqueness, multiplied by the amount of encoded information seems to have also a big impact on memory usage, as

Kraken, CLARK, and CLARK-S require ≥ 80 GB of RAM, partly mitigating the effect of fast k-mer searches through these large indexes. For example, CLARK-S had running times of one order greater than GOTTCHA, while both have a similar pattern matching power of k and CLARK-S might also theoretically encode less information than GOTTCHA (without considering redundancy). The partly reduced non-uniqueness of LMAT with respect to Kraken seems to help better disambiguating reads at the strain level, yielding also the highest recall for all methods, while having a similar precision to Kraken at that level. Nevertheless, this seems to be no more true at higher taxonomic levels, where the use of k-mer frequencies instead of binary values might help better estimating clade abundances and thus filtering the false positive discoveries.

4 Conclusion

We presented a theoretical analysis on the information content used by several state-of-the-art methods in metagenomic sequence classification. In particular, this showed the trade-offs applied by the different types of strategies both in the learning and classification phases, and emphasized several quantitative issues—namely, encoded information, pattern matching power, redundancy, and non-uniqueness—that seem to well correlate with practical factors in the experimental results.

The four quantitative indicators can be indeed useful as a primary consideration when choosing, developing or improving a method for microbial sample analysis, as well as for many bioinformatic applications [7,25].

Through our study, it emerges the need to develop efficient alignment-free methods that overcome such problems, by increasing the encoded information and the pattern matching power of the fastest methods, while efficiently using the available memory and reducing redundancy and non-uniqueness. To this end, one could merge the learning phase and the classification phase, by avoiding the prior extraction of any patterns and deferring this analysis to fast look-up tables, which could allow for a simultaneous, complete comparison of multiple reads to the known reference genomes [10].

References

1. Ames, S.K., Hysom, D.A., Gardner, S.N., Lloyd, G.S., Gokhale, M.B., Allen, J.E.: Scalable metagenomic taxonomy classification using a reference genome database. Bioinformatics **29**(18), 2253–2260 (2013)
2. Breitwieser, F., Baker, D., Salzberg, S.L.: KrakenUniq: confident and fast metagenomics classification using unique k-mer counts. Genome Biol. **19**(1), 198 (2018)
3. Buchfink, B., Xie, C., Huson, D.H.: Fast and sensitive protein alignment using diamond. Nat. Methods **12**(1), 59 (2015)
4. Comin, M., Verzotto, D.: The irredundant class method for remote homology detection of protein sequences. J. Comput. Biol. **18**(12), 1819–1829 (2011)

5. Comin, M., Verzotto, D.: Comparing, ranking and filtering motifs with character classes: application to biological sequences analysis. In: Biological Knowledge Discovery Handbook: Preprocessing, Mining and Postprocessing of Biological Data, chap. 13, pp. 307–332. Wiley (2013)

6. Comin, M., Verzotto, D.: Filtering degenerate patterns with application to protein sequence analysis. Algorithms **6**(2), 352–370 (2013)

7. Comin, M., Verzotto, D.: Beyond fixed-resolution alignment-free measures for mammalian enhancers sequence comparison. IEEE/ACM Trans. Comput. Biol. Bioinf. **11**(4), 628–637 (2014)

8. Comin, M., Verzotto, D.: Alignment-free measures for whole-genome comparison. In: Pattern Recognition in Computational Molecular Biology, chap. 3, pp. 43–64. Wiley (2015)

9. Freitas, T.A.K., Li, P.E., Scholz, M.B., Chain, P.S.: Accurate read-based metagenome characterization using a hierarchical suite of unique signatures. Nucleic Acids Res. **43**(10), e69 (2015)

10. Garofalo, F., Rosone, G., Sciortino, M., Verzotto, D.: The colored longest common prefix array computed via sequential scans. In: Gagie, T., Moffat, A., Navarro, G., Cuadros-Vargas, E. (eds.) SPIRE 2018. LNCS, vol. 11147, pp. 153–167. Springer, Cham (2018). https://doi.org/10.1007/978-3-030-00479-8_13

11. Huson, D.H., Auch, A.F., Qi, J., Schuster, S.C.: Megan analysis of metagenomic data. Genome Res. **17**(3), 377–386 (2007)

12. Lam, T.H., Verzotto, D., Liu, J., Nagarajan, N., et al.: Understanding the microbial basis of body odor in pre-pubescent children and teenagers. Microbiome **6**, 213 (2018)

13. Marchiori, D., Comin, M.: SKraken: fast and sensitive classification of short metagenomic reads based on filtering uninformative k-mers. In: BIOINFORMATICS, pp. 59–67 (2017)

14. McIntyre, A.B., et al.: Comprehensive benchmarking and ensemble approaches for metagenomic classifiers. Genome Biol. **18**(1), 182 (2017)

15. Ounit, R., Lonardi, S.: Higher classification accuracy of short metagenomic reads by discriminative spaced k-mers. In: Pop, M., Touzet, H. (eds.) WABI 2015. LNCS, vol. 9289, pp. 286–295. Springer, Heidelberg (2015). https://doi.org/10.1007/978-3-662-48221-6_21

16. Ounit, R., Wanamaker, S., Close, T.J., Lonardi, S.: Clark: fast and accurate classification of metagenomic and genomic sequences using discriminative k-mers. BMC Genomics **16**(1), 236 (2015)

17. Quince, C., Walker, A.W., Simpson, J.T., Loman, N.J., Segata, N.: Shotgun metagenomics, from sampling to analysis. Nat. Biotechnol. **35**, 833 (2017)

18. Saha, S., Johnson, J., Pal, S., Weinstock, G.M., Rajasekaran, S.: MSC: a metagenomic sequence classification algorithm. Bioinformatics, bty1071 (2019)

19. Teo, A.S., Verzotto, D., Yao, F., Nagarajan, N., Hillmer, A.M.: Single-molecule optical genome mapping of a human HapMap and a colorectal cancer cell line. GigaScience **4**, 65 (2015)

20. Truong, D.T., et al.: Metaphlan2 for enhanced metagenomic taxonomic profiling. Nat. Methods **12**(10), 902 (2015)

21. Vervier, K., Mahé, P., Vert, J.-P.: MetaVW: large-scale machine learning for metagenomics sequence classification. In: Mamitsuka, H. (ed.) Data Mining for Systems Biology. MMB, vol. 1807, pp. 9–20. Springer, New York (2018). https://doi.org/10.1007/978-1-4939-8561-6_2

22. Verzotto, D., Teo, A.S., Hillmer, A.M., Nagarajan, N.: OPTIMA: sensitive and accurate whole-genome alignment of error-prone genomic maps by combinatorial indexing and technology-agnostic statistical analysis. GigaScience **5**, 2 (2016)
23. Verzotto, D., Teo, A.S., Hillmer, A.M., Nagarajan, N.: Index-based map-to-sequence alignment in large eukaryotic genomes (2015)
24. Wood, D.E., Salzberg, S.L.: Kraken: ultrafast metagenomic sequence classification using exact alignments. Genome Biol. **15**(3), R46 (2014)
25. Zielezinski, A., Vinga, S., Almeida, J., Karlowski, W.: Alignment-free sequence comparison: benefits, applications, and tools. Genome Biol. **18**(1), 186 (2017)

mirLSTM: A Deep Sequential Approach to MicroRNA Target Binding Site Prediction

Ahmet Paker[1]([✉]) and Hasan Oğul[1,2]([✉])

[1] Department of Computer Engineering, Başkent University,
06790 Ankara, Turkey
pakerahmett@gmail.com
[2] Faculty of Computer Sciences, Østfold University College,
1757 Halden, Norway
hasan.ogul@hiof.no

Abstract. MicroRNAs (miRNAs) are small and non-coding RNAs of ∼21–23 base length, which play critical role in gene expression. They bind the target mRNAs in the post-transcriptional level and cause translational inhibition or mRNA cleavage. Quick and effective detection of the binding sites of miRNAs is a major problem in bioinformatics. In this study, a deep learning approach based on Long Short Term Memory (LSTM) is developed with the help of an existing duplex sequence model. Compared with four conventional machine learning methods, the proposed LSTM model performs better in terms of the accuracy (ACC), sensitivity, specificity, AUC (Area under the curve) and F1 score. A web-tool is also developed to identify and display the microRNA target sites effectively and quickly.

Keywords: Deep learning · RNN · LSTM · Bioinformatics ·
Sequence Alignment · miRNA · Target prediction · miRNA target site

1 Introduction

MicroRNAs (miRNAs) are responsible for biosynthesis and lysis in tissues within cells. They have ∼21–23 nucleotides length and play critical and gene-regulatory roles in many living organisms. They bind their partial complementary target site and cause cleavage or posttranscriptional repression. They prohibit the genesis of peptides and output proteins [1]. Recent research shown that gene regulation of psychiatric and neurodevelopmental disorders can be observable because of some miRNAs [2, 3]. Although many miRNA target sites are experimentally and computationally determined, a few numbers of them are experimentally validated. Therefore, computational prediction of the function of miRNA targets is a challenging task to support global effort in understanding gene regulation [12, 13].

The main problem is the elucidation of interaction between microRNAs with their target sites. In this process, interaction is done through biological sequences. Recently, a probabilistic model was created to describe the binding preferences between a microRNA sequence and its target site [4]. They proposed a method which is based on a sequential probabilistic model to express miRNA sequences and its related target site.

© Springer Nature Switzerland AG 2019
G. Anderst-Kotsis et al. (Eds.): DEXA 2019 Workshops, CCIS 1062, pp. 38–44, 2019.
https://doi.org/10.1007/978-3-030-27684-3_6

This model converts miRNA-binding site pair (duplex-sequence) to a new sequence. After that, to analyze a new sequence, they used Variable Length Markov Chain (VLMC) [9].

The classification of sequences is a modeling problem that you have a specific input sequence and predicts the target sequence. The difficulty of this problem is that the sequences can vary in length, consist of a large thesaurus of input characters and that the model should examine the long-term context or dependencies between characters in the input sequence. RNN (Recurrent Neural Network) addresses this issue by adding the feedback mechanism which functions as a memory. Thus, the previous inputs in the model are kept in a kind of memory. LSTM expands this idea by creating a short term and long term memory component. As a result, the LSTM model can give successful results in biological sequences which are made of repeating a set of patterns.

For all these reasons, in this work, five different RNN-LSTM based methods have been proposed to solve the miRNA target site prediction problem. Additionally, miRNA and mRNA binding site model is created which includes probabilistic feature representation approach using miRNA - mRNA pairs. Secondly, LSTM Network Model is created. In detail, firstly we used complementary alignment between miRNA and its relevant binding site. Output alignment is shown as a different sequence. With the help of this sequence, every possible match and mismatch is expressed with different alphabet characters. By using these sequences and using the LSTM model which is described in this study, data were analyzed, trained and classified. Mainly five different methods have been performed and they compared with each other. In the first method named as (Method1), we used small size dataset, which is used in [4] to perform with our deep learning model. The reason for doing Method1 is to improve the work of [4]. Therefore, in addition to the data pre-processing method used in [4] study, we attempted to estimate the target sites of the miRNA by a deep learning method, LSTM. In the second method named as (Method2), we used larger size dataset. In this method, without doing any feature representation methods, raw data in [5] is given directly to our deep learning model to predict miRNA target sites. In the third method named as (Method3), we used the same dataset in Method2, additionally, we used data pre-processing and we intended to set up a classification model which is performed a miRNA target site prediction. In the fourth method named as (Method4), there were used DT (Decision Tree) on the same dataset in (Method2). Lastly, in the fifth method named as (Method5), we used DT on the same dataset in (Method1). In conclusion, we compare the results of five methods with using five metrics which are accuracy (ACC), sensitivity, specificity, AUC (Area under the curve) and F1 score.

Finally, a web server is introduced. This application gets miRNA sequence from the user as a text input and shows its all potential binding sites on the related mRNA sequences with red colour. If LSTM Deep Learning model classifies the output as a "Target" according to the input sequences, outputs (Target Sites) are displayed on the screen. The web server is available at https://mirna.atwebpages.com.

2   Related Work

There exist a variety of tools for miRNA target prediction. DeepMirTar [5] is a recent one based on stacked de-noising auto-encoder deep learning method (SdA) to predict human miRNA-targets on the site level. They used three different feature representations to express miRNA targets. They are: High-level expert designed features, Low-Level expert designed features and Raw-data-level designed features. Seed match, sequence composition, free energy, site accessibility, conservation, and hot-encoding are some of the examples of these features. They achieved 93.48% accuracy, 92.35% sensitivity, 94.79% specificity.

TarPmiR [7] developed a random-forest-based approach to predict miRNA target sites. Their method is based on scanning miRNA on related mRNA sequence to get the perfect seed-matching sites. They used six conventional features and seven of their own features, together. Their method calculates the value of these features. As a result, TarPmiR selects the site which has the highest probability as target-site. They achieved 74.46% accuracy, 73.68% sensitivity, 76.56% specificity.

TargetScan v7.0 [8] states that canonical sites are more functional than non-canonical sites to express miRNA binding sites. They extract 14 new features and train the data with multiple linear regression models. They have reached 58.01% accuracy, 60.23% sensitivity, 59.22% specificity.

3   Materials and Methods

3.1   Embedded Vector Representation

The main concept of the embedded word is that every word used in a language can be represented by a set of numeric values (a vector). Embedded words are N-dimensional vectors that try to capture word-meaning and scope in their values.

Firstly, every duplex sequences obtained by the methodology as discussed in Sect. 3.2 converted letter to index. The character "a" converted index 0, Character "b" converted index 1, Character "c" converted index 2, Character "d" converted index 3, Character "q" converted index 4. The Embedded Word method uses the Euclidean distance to find the relationship between similar sequences. Once the dependencies between characters are found, an embedded vector is obtained (Fig. 1).

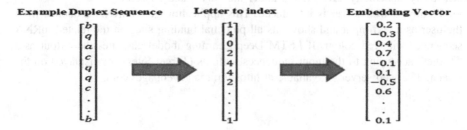

Fig. 1. An example of proposed embedded vector representation

3.2 Proposed LSTM Network

Long short-term memory (LSTM) is a recurrent neural network (RNN) architecture that remembers values at random intervals. Stored values are not changed when learned progress is saved. RNNs allow back and forth connections between neurons.

In our architecture, the first layer is an Embedding layer which will convert our sequences into meaningful numeric embedded vectors as mentioned in Sect. 3.1. In the Embedded layer that uses 32 length vectors to represent each word. The second layer is a Dropout Layer. The main purpose of using this layer is eliminating useless and garbage data. Drop-out percentage is preferred as 20%. The third layer is the LSTM layer with 100 memory units (smart neurons). The fourth layer is a Dropout Layer. Finally, since this is a classification problem, we use a dense output layer with a single neuron and sigmoid activation function to make an estimate of 0 or 1 for two classes in the problem.

Since the binary is a classification problem, log loss is used as a loss function (binary cross-entropy). In addition, as an optimizer "Adam" algorithm is chosen. Adam is an optimization algorithm which updates network weights iteratively in training data. Also to avoid overfitting, in every 5 epoch, validation loss is measured. If validation loss is increased compared to the previous one, early stopping function is activated. As a result, learning is interrupted.

To feed LSTM we use the duplex sequence obtained by complementary alignment of miRNA sequence with a putative target site on mRNA sequence. For every pair of miRNA and target site duplexes are aligned with "Needleman–Wunsch" global complementary alignment algorithm [6]. After that, for every A-U and G-C corresponding nucleotides, different alphabet characters are used. G-U Wobbles are not considered. As a result, it is aimed to decrease the total number of letters in our deep learning model. Each A-U pair is expressed with "a", each U-A pair is expressed with "b", each G-C pair is expressed with "c", each C-G pair is expressed with "d" and all remained pairs are expressed with "q" (Fig. 2).

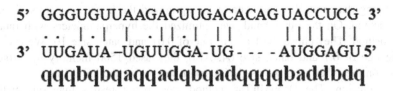

Fig. 2. Example alignment of mRNA binding site (top) and miRNA sequence (middle) converted from a new sequence of duplex formation (bottom).

4 Experiments and Results

4.1 Dataset

In this study, two different datasets were used. First dataset is obtained from Deep-MirTar repository [5]. In the first dataset, 3915 positive data were obtained, 473 of them are taken from mirMark data [10] and 3442 of them are taken from CLASH data [11]. 3905 negative data were produced using mock miRNAs. Totally there are 7820 data. Since there are lots of experimentally validated positive data, this dataset is preferred. Second dataset is taken from [4]. This dataset contains 283 positive and 115 negative miRNA-mRNA duplex sequences. Totally there are 398 data. The purpose of using this dataset is that the DS2 dataset contains experimentally supported positive and negative duplexes. These two datasets are suitable for comparison because they have different sizes. The size of the first data is larger than the second one. Thus, deep learning method is used for the measurement of working performance of different data with different sizes.

4.2   Empirical Results

Firstly, the raw dataset is randomly mixed. After that, LSTM model was evaluated by determining the test split size as 0.1. Hence, 782 randomly selected test data were gathered. 5 metrics were considered to evaluate success criterion: Accuracy (ACC), sensitivity, specificity, AUC and F1 score. According to Table 1, results show that larger size dataset is more convenient than smaller size dataset for used LSTM model. If Method2 and Method3 are compared, pre-processing (Sequence Alignment Technique as mentioned in Sect. 3) and manipulating data gives more consistent and meaningful results. In conclusion, Method3 is more convenient than Method2. In addition, Method3 gives better results than other four basic machine learning methods (TarPmiR, TargetScan v7.0, Method4, Method5) in terms of evaluation metrics which is discussed on Table 1.

In DeepMirTar, they represent the miRNA-mRNA pairs including 750 features. Some of these features are seed match, free energy, sequence composition, site accessibility, etc. On the other hand, our proposed method represents the miRNA-mRNA pairs is based on a probabilistic approach. In the learning phase, they used SdA (Stacked denoising auto-encoder) based on deep neural network. They split dataset 60% training data, 20% validation data, 20% test data. Besides, we split the dataset 90% training data and 10% test data. They optimized the hyperparameters via grid-search method. On the other hand, we optimized the hyperparameters with random search method. They chose the learning rate of 0.01 and batch size 10. We used the learning rate of 0.1 and batch size 64. Also, they used 1500 memory units (smart neurons) on the other hand we used 100 smart neurons in the LSTM layer.

In conclusion, The DeepMirTar method gave better results than our method, because it has a strong optimization of hyperparameters, a more complex deep neural network structure, and a strong input representation to the first layer of the deep network.

Table 1. Result of evaluation metrics for proposed methods and related works on the same dataset

Methods	Accuracy	Sensitivity	Specificity	F1	AUC
Method1	0.725	0.900	0.200	0.830	0.662
Method2	0.803	0.836	0.770	0.807	0.849
Method3	0.840	0.823	0.856	0.835	0.905
Method4	0.813	0.855	0.825	0.813	0.876
Method5	0.700	0.848	0.150	0.824	0.446
DeepMirTar	0.934	0.923	0.947	0.934	0.979
TargetScan v7.0	0.580	0.602	0.592	0.225	0.672
TarPmiR	0.744	0.736	0.765	0.284	0.802

5 Conclusion

This research revealed an LSTM model in addition to miRNA and mRNA target site duplexes model. In this study, a solution to the problem of how miRNA and mRNA sequences interact and bind is proposed. In addition, the problem of connecting miRNA to the target site of mRNA is solved with the help of the proposed LSTM model.

Classical approaches have implemented via window-based scanner over the mRNA sequence to determine the target binding site. On the other hand, deep learning which is based on the LSTM model gives more consistent and significant results because of input sequences lengths are varied also includes lots of different characters. In addition, the model requires a long-term context or dependencies between each different character in the input sequence. LSTM based deep learning model is quite successful to overcome these problems.

For getting better results, seed and non-seed regions can be considered. The number of layers can be increased by using deep learning model. G-U wobbles can be considered at the pre-processing step. Also the free-energy of miRNA- mRNA duplexes can be conceivable.

References

1. Bartel, D.: MicroRNAs: target recognition and regulatory functions. Cell **136**(2), 215–233 (2009)
2. Bartel, D.: MicroRNAs: genomics, biogenesis, mechanism and function. Cell **116**, 281–297 (2004)
3. Xu, B., Hsu, P., Karayiorgou, M., Gogos, J.: MicroRNA dysregulation in neuropsychiatric disorders and cognitive dysfunction. Neurobiol. Dis. **46**(2), 291–301 (2012)
4. Oğul, H., Umu, S., Tuncel, Y., Akkaya, M.: A probabilistic approach to microRNA-target binding. Biochem. Biophys. Res. Commun. **413**(1), 111–115 (2011)
5. Wen, M., Cong, P., Zhang, Z., Lu, H., Li, T.: DeepMirTar: a deep-learning approach for predicting human miRNA targets. Bioinformatics **34**(22), 3781–3787 (2018)

6. Needleman, S.B., Wunsch, C.D.: A general method applicable to the search for similarities in the amino acid sequences of two proteins. J. Mol. Biol. **48**, 443–453 (1970)
7. Ding, J., Li, X., Hu, H.: TarPmiR: a new approach for microRNA target site prediction. Bioinformatics **32**, 2768–2775 (2016)
8. Agarwal, V., Bell, G., Nam, J., Bartel, D.: Predicting effective microRNA target sites in mammalian mRNAs. eLife **4**, e05005 (2015)
9. Ron, D., Singer, Y., Tishby, N.: The power of amnesia: learning probabilistic automata with variable memory length. Mach. Learn. **25**, 117–149 (1996)
10. Menor, M., et al.: mirMark: a site-level and UTR-level classifier for miRNA target prediction. Genome Biol. **15**, 500 (2014)
11. Helwak, A., et al.: Mapping the human miRNA interactome by CLASH reveals frequent noncanonical binding. Cell **153**, 654–665 (2013)
12. Dede, D., Oğul, H.: TriClust: a tool for cross-species analysis of gene regulation. Mol. Inf. **33**(5), 382–387
13. Oğul, H., Akkaya, M.S.: Data integration in functional analysis of microRNAs. Curr. Bioinf. **6**, 462–472 (2011)

K – Means Based One-Class SVM Classifier

Loai Abedalla[1]([⊠]), Murad Badarna[3], Waleed Khalifa[2],
and Malik Yousef[4]

[1] Department of Information Systems, Yezreel Valley Academic College,
Yezreel Valley, Israel
loaia@yvc.ac.il
[2] Computer Science, The College of Sakhnin, Sakhnin, Israel
[3] Department of Information Systems, University of Haifa, Haifa, Israel
[4] Department of Information Systems, Zefat Academic College, Safed, Israel

Abstract. The application of one-class machine learning is gaining attention in the computational biology community. Many biological cases can be considered as multi one-class classification problem. Examples include the classification of multiple cancer types, protein fold recognition and, molecular classification of multiple tumor types. In all of those cases the real world appropriately characterized negative cases or outliers are impractical to be achieved and the positive cases might be consists from different clusters which in turn might reveal to accuracy degradation. In this paper, we present multi-one-class classifier to deal with this problem. The key point of our classification method is to run a clustering algorithm such as the well-known k-means over the positive cases and then building up a classifier for every cluster separately. For a given new example, we apply all the generated classifiers. If it rejected by all of those classifiers, the given example will be considered as a negative case, otherwise it is a positive case.

Keywords: One class SVM · Clustering based classification · K-means · Ensemble clustering

1 Introduction

The one-class approach in machine learning has been receiving more attention particularly for solving problems where the negative class is not well defined [1–5]; moreover, the one class approach has been successfully applied in various fields including text mining [6], functional Magnetic Resonance Imaging (fMRI) [7], signature verification [8] and miRNA gene and target discovery [5, 9].

One and two class classification methods can both give useful classification accuracies. The advantage of one-class methods is that they do not require any additional effort for choosing the best way of generating the negative class. Some data is composed from multiple category or classes. For such a data, special methods are required and the one class gives insufficient results.

Here, we present a new approach that devise the positive class into sub-groups and then builds a classifier (one-class) for each sub-group.

© Springer Nature Switzerland AG 2019
G. Anderst-Kotsis et al. (Eds.): DEXA 2019 Workshops, CCIS 1062, pp. 45–53, 2019.
https://doi.org/10.1007/978-3-030-27684-3_7

The key point of our classification method is to run the k-means clustering algorithm over the positive cases and then building up a classifier for every cluster separately. For a given new sample the algorithm applies all the generated classifiers. If one of those classifiers classifies the given sample as a positive sample then it will be considered as a positive sample, otherwise it is a negative sample.

The problem of how to implement a multi-class classifier by an ensemble of one-class classifiers has attracted much attention in the machine-learning community [10, 11]. Tax et al. [12] proposed a method for combining different one-class classifiers, which improves dramatically the performance and the robustness of the classification, to the handwritten digit recognition problem. Similar method investigated by Lai et al. [13] to the problem of image retrieval problem. They reported that combining multi SVM based classifiers improves the retrieval precision.

Several methods for handling missing feature values which are based on combining one-class classifiers have been suggested by Juszczak et al. [14]. They indicated that their methods are more flexible and more robust to small sample size problems than the standard methods such as the linear programming dissimilarity-data description. Ban et al. [15] suggested using multiple one-class classifiers which can deal with the feature space and the nonlinear classification problem. They trained the multi one-class classifiers on each class and then extracted a decision function which based on minimum distance rules. Their experiments showed that the proposed method outperforms the OC-SVM.

Lyu and Fraid [16] provided a multi one-class SVMs which combines a beforehand clustering process for detecting hidden (Steganographic) messages in digital images. They show how a multi one-class SVM greatly simplifies the training stage of the classifiers. Their results show that even though the overall detection improves with an increasing number of hyper-spheres (i.e., the clusters), the false-positive rate begins to increase considerably with the increases of the number of the hyper-spheres. Recently, Menahem et al. [17] described a different multiple one-class classification approach called TUPSO which combines multi one-class classifiers via meta-classifier. They showed that TUPSO outperforms existing methods such as the OC-SVM.

Much works on multi one-class classification are existed concerning the computational biology community. Spinosa et al. [18] suggested a multi one-class classification technique to detect novelty in gene expression data. They combined different one-class classifiers such as OC-Kmeans and OC-KNN. For a given example, each classifier votes 'positive' or 'outlier'. The final decision is taken according to the majority votes of all classifiers. Results showed that the robustness of the classification is increased because it takes into account different classifiers such that each one judges the examples in different point of view. Recently, similar approach provided by Zhang et al. [19] for the avian influenza outbreak classification problem.

Our new approach distinguished itself from those predecessors in the way that it first clusters the positive data into groups before the classification process. This pre-processing prevents the drawback of using only a single hyper-sphere (i.e., the one generated by the one-class classifier) which may not provide a particularly compact support for the training data. Our experiments show that our approach significantly improves the accuracy of the classification.

The rest of this paper is organized as follows: Sect. 2 describes necessary preliminaries. Our multi one-class approach is described in Sect. 3 and evaluated in Sect. 4. Our main conclusions and future work can be found in Sect. 5.

2   Methods

2.1   One-Class Methods

In general, a binary learning (two-class) approach to miRNA discovery considers both positive (miRNA) and negative (non-miRNA) classes by providing examples for the two-classes to a learning algorithm in order to build a classifier that will attempt to discriminate between them. The most common term for this kind of learning is *supervised learning* where the labels of the two-classes are known beforehand.

One-class uses only the information for the target class (positive class) building a classifier which is able to recognize the examples belonging to its target and rejecting others as outliers. Among the many classification algorithms available, we chose four one-class algorithms to compare for miRNA discovery. We give a brief description of each one-class classifier and we refer the references [20, 21] for additional details including a description of *parameters* and thresholds. The LIBSVM library [22] was used as implementation of the SVM (one-class using the RBF kernel function) and the DDtools [23] for the other one-class methods. The WEKA software [24] was used as implementation of the two-class classifiers.

2.2   One-Class Support Vector Machines (OC-SVM)

Support Vector Machines (SVMs) are a learning machine developed as a two-class approach [25, 26]. The use of one-class SVM was originally suggested by [21]. One-class SVM is an algorithmic method that produces a prediction function trained to "capture" most of the training data. For that purpose a kernel function is used to map the data into a feature space where the SVM is employed to find the hyper-plane with maximum margin from the origin of the feature space. In this use, the margin to be maximized between the two classes (in two-class SVM) becomes the distance between the origin and the support vectors which define the boundaries of the surrounding circle, (or hyper-sphere in high-dimensional space) which encloses the single class.

2.3   One-Class Gaussian (OC-Gaussian)

The Gaussian model is considered as a density estimation model. The assumption is that the target samples form a multivariate normal distribution, therefore for a given test sample z in n-dimensional space, the probability density function can be calculated as:

$$p(z) = \frac{1}{(2\pi)^{\frac{n}{2}}|\Sigma|^{\frac{1}{2}}} e^{-\frac{1}{2}(z-\mu)^T \Sigma^{-1}(z-\mu)} \qquad (1)$$

where μ and Σ are the mean and covariance matrix of the target class estimated from the training samples.

2.4   One-Class Kmeans (OC-Kmeans)

Kmeans is a simple and well-known unsupervised machine learning algorithm used in order to partition the data into k clusters. Using the OC-Kmeans we describe the data as k clusters, or more specifically as k centroids, one derived from each cluster. For a new sample, z, the distance $d(z)$ is calculated as the minimum distance to each centroid. Then based on a user specified threshold, the classification decision is made. If $d(z)$ is less than the threshold the new sample belongs to the target class, otherwise it is rejected.

2.5   One-Class K-Nearest Neighbor (OC-KNN)

The one-class nearest neighbor classifier (OC-KNN) is a modification of the known two-class nearest neighbor classifier which learns from positive examples only. The algorithm operates by storing all the training examples as its model, then for a given test example z, the distance to its nearest neighbor y ($y = NN(z)$) is calculated as $d(z,y)$. The new sample belongs to the target class when:

$$\frac{d(z,y)}{d(y,NN(y))} < \delta \qquad (2)$$

where $NN(y)$ is the nearest neighbor of y, in other words, it is the nearest neighbor of the nearest neighbor of z. The default value of δ is 1. The average distance of the k nearest neighbors is considered for the OC-KNN implementation.

3   Multi One-Class Classifiers

For a given data, the positive class might be consisting of different clusters (See Fig. 1). If we try to find the small hyper-sphere that contains all the data points belongs to the positive class, then the negative data points will be a part of the hyper-sphere yielding in low performance.

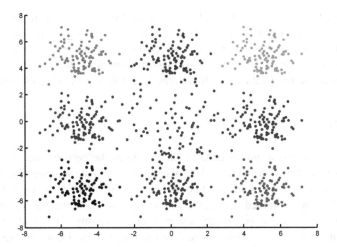

Fig. 1. The positive class consists of 4 sub-groups. The negative class is in blue color. Each cluster has a different color (pink, green, black, and red). (Color figure online)

To alleviate this type of data we propose the Multi One-Class Classifiers that works with small groups of the positive data. Our approach based on the OC-SVM that builds a hypersphere for a given positive data and we aim that working with compacted groups is better that work with wide group.

As a result, we run clustering to identify the similar data points. The main idea of our approach is to cluster the positive data into several clusters using the k-means clustering algorithm and then build up a classifier from each cluster using the OC-SVM as can be seen in Fig. 2.

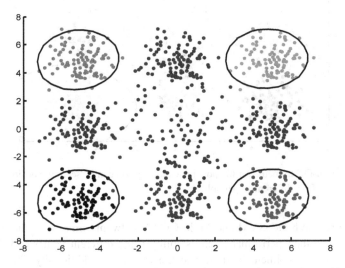

Fig. 2. The Multi-OC trained over the positive examples. As can be seen, the positive examples are classified into four different hyper-spheres.

For a given new example, we apply all the classifiers. If all of them reject it then it considers as negative example; otherwise it considers as belongs to the one-class data. In this work, without loss of generality, we choose to evaluate our Multi One-Class approach using the OC-SVM as the basis classification algorithm and the standard K-means as the clustering method. We leave other possible combinations for future work.

4   Results

The first experiment is a typical one, which illustrates the performance of Multi-OC-SVM versus that of the normal OC-SVM over a synthetic data set (See Fig. 1). In each experiment, the data set is split into two subsets, one for training and one for testing. Both algorithms were trained using 80% of the positive class and the remaining 20% together with all the negative examples were used for testing. The positive examples are divided into four clusters beforehand. We used the standard k-means algorithm for this purpose.

As can be seen from Fig. 3, executing the OC-SVM method reveals into poor classification. This is to be expected since OC-SVM classifies part of the positive examples as negative ones (or as outliers). Figure 4 depicts the classification done by the Multi-OC-SVMs. As can be seen, obviously Multi-OC-SVM outperforms OC-SVM.

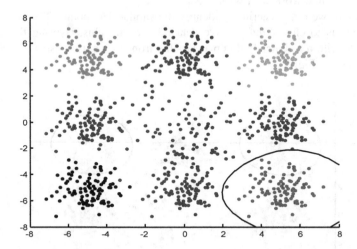

Fig. 3. The Multi-OC trained over the positive examples. As can be seen, the positive examples are classified into four different hyper-spheres.

Next we ran both OC-SVM and Multi-OC-SVM twenty times over the data set in order to evaluate their performance and stability using different values of k. The data set used for this experiment is similar to the first one. Additionally, the Matthews Correlation Coefficient (MCC) (see [27] for more details) measurement is used to take into account both over-prediction and under-prediction in imbalanced data sets. Looking at Fig. 4, one can see that the accuracy of Multi-OC-SVM is far better than that of OC-SVM. Furthermore, it shows that Multi-OC-SVM is stable over different number of clusters.

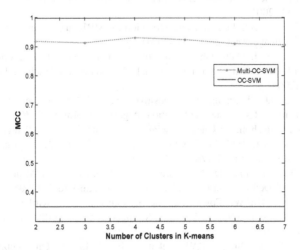

Fig. 4. The accuracy of OC-SVM and Multi-OC-SVM as a function of the number of clusters generated by the Kmeans clustering algorithm.

5 Conclusion

The current results show that it is possible to build up a multi one-class classifier with a combined clustering beforehand process based only on positive examples yielding a reasonable performance.

Further research can proceed in several interesting directions. First, the suitability of the framework of our approach to different data types can be investigated. Second, it would be interesting to apply our approach to other types of classifiers and to more robust clustering methods such as Mean-Shift [28]. Lastly, it would be interesting to test our approach on more real data-sets and real problems.

References

1. Kowalczyk, A., Raskutti, B.: One class SVM for yeast regulation prediction. SIGKDD Explor. **4**(2), 99–100 (2002)
2. Spinosa, E.J., Carvalho, A.C.: Support vector machines for novel class detection in Bioinformatics. Genet. Mol. Res. (GMR) **4**(3), 608–615 (2005)
3. Crammer, K., Chechik, G.: A needle in a haystack: local one-class optimization. In: Proceedings of the Twenty-First International Conference on Machine Learning (ICML) (2004)

4. Gupta, G., Ghosh, J.: Robust one-class clustering using hybrid global and local search. In: Proceedings of the 22nd International Conference on Machine Learning. ACM Press, Bonn (2005)
5. Yousef, M., Najami, N., Khalifa, W.: A comparison study between one-class and two-class machine learning for MicroRNA target detection. J. Biomed. Sci. Eng. **3**, 247 (2010)
6. Manevitz, L.M., Yousef, M.: One-class SVMs for document classification. J. Mach. Learn. Res. **2**, 139–154 (2001)
7. Thirion, B., Faugeras, O.: Feature characterization in fMRI data: the information bottleneck approach. Med. Image Anal. **8**(4), 403 (2004)
8. Koppel, M., Schler, J.: Authorship verification as a one-class classification problem. In: Proceedings of the Twenty-First International Conference on Machine Learning. ACM Press, Banff (2004)
9. Yousef, M., et al.: Learning from positive examples when the negative class is undetermined- microRNA gene identification. Algorithms Mol. Biol. **3**(1), 2 (2008)
10. AbedAllah, L., Shimshoni, I.: k nearest neighbor using ensemble clustering. In: Cuzzocrea, A., Dayal, U. (eds.) DaWaK 2012. LNCS, vol. 7448, pp. 265–278. Springer, Heidelberg (2012). https://doi.org/10.1007/978-3-642-32584-7_22
11. Landgrebea, T.C., Paclika, D.M., Andrew, R.P.: One-Class and Multi-Class Classifier Combining for Ill-Defined Problems. Elsevier Science, Amsterdam (2005)
12. Tax, D.M.J., Duin, R.P.W.: Combining one-class classifiers. In: Kittler, J., Roli, F. (eds.) MCS 2001. LNCS, vol. 2096, pp. 299–308. Springer, Heidelberg (2001). https://doi.org/10. 1007/3-540-48219-9_30
13. Lai, C., Tax, D.M.J., Duin, R.P.W., Pękalska, E., Paclík, P.: On combining one-class classifiers for image database retrieval. In: Roli, F., Kittler, J. (eds.) MCS 2002. LNCS, vol. 2364, pp. 212–221. Springer, Heidelberg (2002). https://doi.org/10.1007/3-540-45428-4_21
14. Juszczak, P., Duin, R.P.W.: Combining one-class classifiers to classify missing data. In: Roli, F., Kittler, J., Windeatt, T. (eds.) MCS 2004. LNCS, vol. 3077, pp. 92–101. Springer, Heidelberg (2004). https://doi.org/10.1007/978-3-540-25966-4_9
15. Ban, T., Abe, S.: Implementing multi-class classifiers by one-class classification methods. In: International Joint Conference on Neural Networks, IJCNN 2006 (2006)
16. Lyu, S., Farid, H.: Steganalysis using color wavelet statistics and one-class support vector machines. In: SPIE Symposium on Electronic Imaging, pp. 35–45 (2004)
17. Menahem, E., Rokach, L., Elovici, Y.: Combining One-Class Classifiers via Meta-learning. CoRR, abs/1112.5246 (2011)
18. Spinosa, E.J., de Carvalho, A.C.P.L.F.: Combining one-class classifiers for robust novelty detection in gene expression data. In: Setubal, J.C., Verjovski-Almeida, S. (eds.) BSB 2005. LNCS, vol. 3594, pp. 54–64. Springer, Heidelberg (2005). https://doi.org/10.1007/ 11532323_7
19. Zhang, J., Lu, J., Zhang, G.: Combining one class classification models for avian influenza outbreaks. In: Computational Intelligence in Multicriteria Decision-Making (MDCM), pp. 190–196. IEEE (2011)
20. Tax, D.M.J.: One-class classification; concept-learning in the absence of counter-examples. Delft University of Technology, June 2001
21. Schölkopf, B., et al.: Estimating the support of a high-dimensional distribution. Neural Comput. **13**(7), 1443–1471 (2001)
22. Chang, C.-C., Lin, C.-J.: LIBSVM: a library for support vector machines. ACM Trans. Intell. Syst. Technol. (TIST) **2**, 27 (2001)
23. Tax, D.M.J.: DDtools, the Data Description Toolbox for Matlab. Delft University of Technology (2005)

24. Witten, I.H., Frank, E.: Data Mining: Practical Machine Learning Tools and Techniques, 2nd edn. Morgan Kaufmann, San Francisco (2005)
25. Schölkopf, B., Burges, C.J.C., Smola, A.J.: Advances in Kernel Methods. MIT Press, Cambridge (1999)
26. Vapnik, V.: The Nature of Statistical Learning Theory. Springer, Heidelberg (1995). https://doi.org/10.1007/978-1-4757-3264-1
27. Matthews, B.W.: Comparison of the predicted and observed secondary structure of T4 phage lysozyme. Biochim. Biophys. Acta 405(2), 442–451 (1975)
28. Comaniciu, D., Meer, P.: Mean shift: a robust approach toward feature space analysis. IEEE Trans. Pattern Anal. Mach. Intell. 24(5), 603–619 (2002)

Molecular Subtyping in Human Disease Using the Paraclique Algorithm

Ronald D. Hagan[1](✉) and Michael A. Langston[2]

[1] BAE Systems, Burlington, MA 01803, USA
ron.hagan@baesystems.com
[2] Department of Electrical Engineering and Computer Science,
University of Tennessee, Knoxville, TN 37996, USA
langston@tennessee.edu

Abstract. Recent discoveries of distinct molecular subtypes have led to remarkable advances in treatment for a variety of diseases. While subtyping via unsupervised clustering has received a great deal of interest, most methods rely on basic statistical or machine learning methods. In this paper we discuss a method based on the paraclique algorithm, and demonstrate its potential effectiveness through testing on four sets of publicly available gene expression microarray data.

Keywords: Molecular subtyping · Paraclique · Graph algorithms

1 Introduction

It has long been established that many families of disease exhibit a wide range of heterogeneity. This is especially true in many cancers. Lung cancers, for example, fall into two overall types based on histological characteristics: small cell lung cancer (SCLC), and non-small cell lung cancer (NSCLC). In addition, non-small cell lung cancer can be further stratified based on pathological characteristics into four principal types: adenocarcinoma, squamous cell carcinoma, large cell carcinoma, and NSCLC not otherwise specified (NOS). While histological classification remains crucial, significant advances in the treatment of NSCLC over the last decade have centered around the development of therapies targeting subtypes at the molecular level, such as genetic mutations. In particular, therapies targeting alterations in the epidermal growth factor receptor (EGFR) and anaplastic lymphoma kinase (ALK) genes using kinase inhibitors have produced dramatic improvements in outcome for patients in the underlying subgroups [1, 2]. In addition to providing new paths for therapy, advances in molecular subtyping are allowing practitioners to avoid needless high-risk therapies. For example, studies have identified gene expression signatures for chemo-resistance in both acute myeloid leukemia and breast cancer [3, 4]. In each, the key development driving these advances is the successful identification of molecular subtypes.

While techniques grounded in graph theory have been used to great effect in the pursuit of genetic biomarkers for human disease and the discovery of novel gene networks, little work has been done in trying to extend such tools to subtyping. The existing research has primarily centered on the use of basic statistical or machine

© Springer Nature Switzerland AG 2019
G. Anderst-Kotsis et al. (Eds.): DEXA 2019 Workshops, CCIS 1062, pp. 54–58, 2019.
https://doi.org/10.1007/978-3-030-27684-3_8

learning clustering methods such as k-means, latent variable models, or mixture models. In this paper, we describe a method for the discovery of putative subtypes based on molecular signatures using the paraclique algorithm. While our method is general and applies easily to other types of data such as protein, metabolite, or DNA methylation profiles, we focus our experimentation in this work on gene expression data.

2 Algorithmic Workflow

Clique-centric methods have long been used for the discovery and modeling of coherent networks. Unfortunately, clique finders can be inherently prone to high false negative rates. Indeed, an entire clique will be missed if even a single edge is lost. The paraclique algorithm was introduced in [5], largely to ameliorate such difficulties posed by noisy data. While we will not delve into the algorithmic details of paraclique construction here (see [6]), the basic idea is to begin with a maximum clique as a core, then expand it by glomming onto vertices adjacent to all but some allowable number of core vertices.

In an effort to limit the effects of confounding factors, we employ an initial filtering step as a simple form of feature selection. FDR corrected p-values for differential expression of genes between case and control sets are calculated, and only those with p-values less than 0.1 are retained. Our aim is that this technique will tend to limit our attention to those genes of interest only in the case group. After all, we are not interested in subgroups based on age, ethnicity, hair color and so forth. These p-values are calculated using the EntropyExplorer R package [7]. After filtering is complete, we transpose the data tables and calculate the pairwise Pearson correlation coefficients between samples across the expression levels. By thresholding the correlation matrix, we create an unweighted graph with vertices representing individual samples and edges denoting pairs of samples correlated above the threshold (in absolute value). Once the case graph is thereby constructed, we run the paraclique algorithm to extract dense subgraphs of patients representing putative subtypes. We note that this approach provides inherent quality control as well: given a patient population of even modest size, any patient who forms a singleton cluster merits additional scrutiny as a possible disease misclassification or other form of outlier.

3 Experimental Results

We chose four well-known and widely-studied diseases (Asthma, Breast Cancer, Chronic Lymphocytic Leukemia (CLL) and Colorectal Cancer), and applied our methods to the corresponding publicly available gene expression data obtained from the Gene Expression Omnibus (GEO). Our investigation was focused on two guiding questions. First, is the method capable of reliably identifying putative subtypes? And second, is there any other evidence confirming or at least supporting these subtypes as being biologically relevant to the associated disease?

We found the answer to the first question to be an unequivocal "yes." As summarized in Table 1, we found putative subtypes in each of our datasets.

Table 1. Summary of diseases tested and size of putative subgroups identified.

Disease	GEO ID	Case	Control	Size of subgroups
Asthma	GSE4302	42	28	31,8,3
Breast Cancer	GSE10810	31	27	22,5
CLL (Leukemia)	GSE8835	24	12	4,18
Colorectal Cancer	GSE9348	70	12	63,5

The second question seems considerably more nuanced and complex. For it, we followed a two-step approach. To begin, we calculated GO enrichments for the top 100 differentially expressed genes, as well as their associated enrichment p-values. The results, along with enriched GO categories, are summarized in Table 2. Note that, in each example, we found significant evidence for biological significance among the gene sets separating our samples into subgroups with enrichment p-values ranging from 1.1E–4 for the asthma data to 2.82E–28 for colorectal cancer.

Table 2. GO enrichment p-values of 100 most differentially expressed genes across subgroups

Data set	Category	p-value
Asthma GSE4302	Oxireductase	1.1E–4
Breast Cancer GSE10810	Secreted	1.0E–13
Leukemia GSE8835	Mhc ii	2.4E–15
Colorectal Cancer GSE9348	Translational elongation	2.8E–28

We then performed a literature search to check the top gene lists for involvement in known disease subtypes. As discussed below, we were able to find strong empirical evidence to support having successfully identified such subtypes for each dataset.

Asthma. The GEO series GSE4302 consists of expression data derived from epithelial airway brushings taken from 42 asthmatics and a control group consisting of 28 healthy subjects and 16 smokers. For our analysis, we used only the healthy subjects as control, discarding the smokers. Data is derived from a microarray analysis using the Affymetrix Human Genome U133 Plus 2.0 Array. Our initial filtering step reduced the original 54,676 probes to a set of 2322 having FDR corrected p-values less than 0.1 for differential expression. With the reduced set of variables and a threshold of 0.93, our method produced three putative subgroups of size 31, 8 and 3. The most differentially expressed genes between the case and control included CLCA1, periostin and ovalbumin. All three of these genes were reported in [8] to be markers of a Th2-high endotype of asthma.

Breast Cancer. GSE10810 contains gene expression data for 31 tumor samples and a control set of 27 healthy tissue samples. The platform used for the study was again the Affymetrix Human Genome U133 Plus 2.0 Array, although only data for 18,382 probes was provided. The number of probes was reduced to 11,531 with false discovery rate adjusted p-values less than 0.1 for differential expression. Using a threshold of 0.8, our tools produced two putative subgroups of size 22 and 5. The most differentially expressed genes between case and control include SLC39A6, S100a4, AGR3, Cd24 and epcam. All these genes have been reported in the literature as being markers for different phenotypes of breast cancer [9, 10].

CLL (Chronic Lymphocytic Leukemia). The dataset GSE8835 is made up of 24 CD4 cell samples from CLL patients and a control group of 12 CD4 cell samples from healthy, age matched donors. The Affymetrix Human Genome U133A Array with 22,283 probes was used. Filtering reduced the probes to a set of 1338 having p-values less than 0.1 (again FDR corrected). A threshold of 0.8 produced two subgroups of size 4 and 18. The most differentially expressed genes between case and control included ZAP-70, previously identified as the best discriminator of Ig-mutated versus Ig-unmutated CLL [11].

Colorectal Cancer. GSE9348 consists of gene expression data derived from tumors of 70 patients and tissue from 12 healthy controls. The samples are age and ethnicity matched. The study utilized the Affymetrix U133 Plus 2.0 array. Our filtering reduced the 54675 probes in the original set to 22968 with FDR corrected p-values less than 0.1 for differential expression. With a threshold of 0.87, our tools produced two paracliques representing putative subgroups in the tumor samples of size 63 and 5. The most differentially expressed genes between case and control included Cd24, identified as a prognostic marker for colorectal cancer [12] as well as OLFM4, indicated as a marker for tumor differentiation and progression [13].

4 Summary

We described a method based on the paraclique algorithm that can identify putative subtypes by separating samples based on signatures in their molecular profiles. While our tools are easily extensible to other types of data, we focused our efforts on gene expression data as an early proof of concept. We applied these methods to four sets of publicly available data obtained from the Gene Expression Omnibus, identified putative subtypes for each set, and found literature support for the subtypes thus identified. The results of this brief study suggest a strong potential for the utility of this overall approach in the reinforcement of known subtypes, in the discovery of novel subtypes, and in the identification and elimination of misclassified data and other sorts of outliers.

References

1. Mok, T.S., et al.: Gefitinib or carboplatin-paclitaxel in pulmonary adenocarcinoma. N. Engl. J. Med. **361**(10), 947–957 (2009)
2. Shaw, A.T., et al.: Crizotinib versus chemotherapy in advanced ALK-positive lung cancer. N. Engl. J. Med. **368**(25), 2385–2394 (2013)
3. Leith, C.P., et al.: Acute myeloid leukemia in the elderly: assessment of multidrug resistance (MDR1) and cytogenetics distinguishes biologic subgroups with remarkably distinct responses to standard chemotherapy. A Southwest Oncology Group study. Blood **89**(9), 3323–3329 (1997)
4. Balko, J.M., et al.: Profiling of residual breast cancers after neoadjuvant chemotherapy identifies DUSP4 deficiency as a mechanism of drug resistance. Nat. Med. **18**(7), 1052–1059 (2012)
5. Chesler, E.J., Langston, M.A.: Combinatorial genetic regulatory network analysis tools for high throughput transcriptomic data. In: Eskin, E., Ideker, T., Raphael, B., Workman, C. (eds.) RRG/RSB -2005. LNCS, vol. 4023, pp. 150–165. Springer, Heidelberg (2007). https://doi.org/10.1007/978-3-540-48540-7_13
6. Hagan, R.D., Langston, M.A., Wang, K.: Lower Bounds on Paraclique Density. Discret. Appl. Math. **204**, 208–212 (2016)
7. Wang, K., Phillips, C.A., Saxton, A.M., Langston, M.A.: EntropyExplorer: an R package for computing and comparing differential Shannon entropy, differential coefficient of variation and differential expression. BMC Res. Notes **8**, 832 (2015)
8. Woodruff, P.G.: Subtypes of asthma defined by epithelial cell expression of messenger RNA and microRNA. Annal. Am. Thorac. Soc. **10**(Suppl.), S186–S189 (2013)
9. Srour, N., Reymond, M.A., Steinert, R.: Lost in translation? a systematic database of gene expression in breast cancer. Pathobiology **75**(2), 112–118 (2008)
10. de Silva Rudland, S., et al.: Association of S100A4 and osteopontin with specific prognostic factors and survival of patients with minimally invasive breast cancer. Clin. Cancer Res. **12**(4), 1192–1200 (2006)
11. Wiestner, A., et al.: ZAP-70 expression identifies a chronic lymphocytic leukemia subtype with unmutated immunoglobulin genes, inferior clinical outcome, and distinct gene expression profile. Blood **101**(12), 4944–4951 (2003)
12. Belov, L., Zhou, J., Christopherson, R.I.: Cell surface markers in colorectal cancer prognosis. Int. J. Mol. Sci. **12**(1), 78–113 (2010)
13. Besson, D., et al.: A quantitative proteomic approach of the different stages of colorectal cancer establishes OLFM4 as a new nonmetastatic tumor marker. Mol. Cell. Proteomics **10**(12), M111.009712 (2011)

Speeding-Up the Dynamic Programming Procedure for the Edit Distance of Two Strings

Giuseppe Lancia[1(✉)] and Marcello Dalpasso[2]

[1] Dipartimento di Scienze Matematiche, Informatiche e Fisiche, University of Udine,
Via delle Scienze 206, 33100 Udine, Italy
giuseppe.lancia@uniud.it

[2] Dipartimento di Ingegneria dell'Informazione, University of Padova,
Via Gradenigo 6/A, 35131 Padova, Italy
marcello.dalpasso@unipd.it

Abstract. We describe a way to compute the edit distance of two strings without having to fill the whole dynamic programming (DP) matrix, through a sequence of increasing guesses on the edit distance. If the strings share a certain degree of similarity, the edit distance can be quite smaller than the value of non-optimal solutions, and a large fraction (up to 80–90%) of the DP matrix cells do not need to be computed. Including the method's overhead, this translates into a speedup factor from 3× up to 30× in the time needed to find the optimal solution for strings of length about 20,000.

1 Problem and Notation

Let Σ be an alphabet and s', s'' be two strings over Σ. We can always turn s' into s'' through a sequence of three basic operations:

- *Deletion* of a symbol σ of s': cost $\mathtt{del}(\sigma)$.
- *Insertion* of a symbol τ of s'' into s': cost $\mathtt{ins}(\tau)$.
- *Substitution* of a symbol σ of s' with a symbol $\tau \neq \sigma$ of s'': cost $\mathtt{sub}(\sigma, \tau)$.

The cost of the sequence is the sum of the costs of the individual operations. The *Edit Distance* problem calls for computing a sequence of operations of minimum cost, called the *edit distance* of s' and s'', here $d(s', s'')$. We assume the costs are positive (so that $d(s', s'') = 0 \Rightarrow s' = s''$) and $s' \neq s''$, so that $d(s', s'') > 0$. For convenience, we define $\mathtt{sub}(\sigma, \sigma) := 0$ for all σ.

Usually the costs are represented in the form of a *substitution matrix*, i.e., a square matrix of order $|\Sigma| + 1$, with 0's on the diagonal. The last row and column of the substitution matrix are associated to an extra symbol '-', called *gap*, and contain the costs of insertions and deletions.

Some substitution matrices (here called *simple*) have the property that all insertions and deletions have the same cost (the *indel* cost, IND) and all substitutions have the same cost (SUB). For instance the standard substitution matrix

G. Anderst-Kotsis et al. (Eds.): DEXA 2019 Workshops, CCIS 1062, pp. 59–66, 2019.
https://doi.org/10.1007/978-3-030-27684-3_9

for the edit distance (called `ones.mat` here) is simple, with $\texttt{IND} = \texttt{SUB} = 1$. Another popular simple substitution matrix is used for DNA comparison and has $\texttt{SUB} = 15$ and $\texttt{IND} = 10$: we call this matrix `dna.mat`. In general, in a substitution matrix we can assume that $\texttt{sub}(\sigma, \tau) < \texttt{del}(\sigma) + \texttt{ins}(\tau)$ (otherwise no substitution would ever be made since it can be better mimicked by a deletion followed by an insertion). For a simple matrix this becomes $\texttt{SUB} < 2\,\texttt{IND}$.

The edit distance problem, first investigated by Levenshtein in [5] (and therefore also known as *Levenshtein distance*), is a classic of string-related problems [3], with main applications in the field of bioinformatics [4]. In this context, Σ is either the alphabet of the 4 nucleotides or of the 20 amino acids, and the problem is solved to determine the similarity between two genomic sequences (i.e., to *align* them in the best possible way). It is effectively solved via a $\Theta(nm)$ dynamic programming (DP) procedure [2,6], where $n = |s'|$ and $m = |s''|$.

The DP procedure is a two-level nested `for` cycle which fills a table $\texttt{P}[\cdot, \cdot]$ of $n + 1$ rows and $m + 1$ columns. At the end, the value of a generic entry $\texttt{P}[i,j]$ with $0 \leq i \leq n$ and $0 \leq j \leq m$ is equal to the edit distance between the prefixes $s'[1, \ldots, i]$ and $s''[1, \ldots, j]$. The standard code is like this (we assume that $\texttt{P}[x,y]$ returns $+\infty$ if either $x < 0$ or $y < 0$):

```
P[0,0]:=0
for  j := 1 to m do P[0, j] := P[0, j − 1] + ins(s″[j])
for  i := 1 to n do
    for  j := 0 to m do
        P[i,j] := min{P[i, j − 1] + ins(s″[j]), P[i − 1, j] + del(s′[i]),
                P[i − 1, j − 1] + sub(s′[i], s″[j])}
```

Hence, $d(s', s'') = \texttt{P}[n, m]$. The corresponding optimal sequence of operations can be retrieved, for instance, by starting at the cell $v := (n, m)$ and determining which cell, among the candidates $(n, m − 1)$, $(n − 1, m − 1)$ and $(n − 1, m)$, is responsible for the value of v; then, set v to be such a cell and proceed backtracking in the matrix until $v = (0, 0)$. We say that the cells thus touched are on an *optimal path* P^* from $(0, 0)$ to (n, m).

Our Goal. In retrospective, once $\texttt{P}[\cdot, \cdot]$ has been filled and we backtrack along the optimal path, we see that there are some cells of $\texttt{P}[\cdot, \cdot]$ whose value is so large that they could have never been on an optimal path but they have been computed nonetheless. Ideally, we would have liked to fill $\texttt{P}[\cdot, \cdot]$ only in the cells of P^*, but the knowledge of these cells needs the knowledge of some other cells, adjacent to them, and these in turn need some other cells, etc., so that it might look impossible to avoid computing the value of some of the $(n+1)(m+1)$ cells, since we cannot exclude that any particular cell could belong to P^* or have an effect on P^*.

Indeed, as we will show, this is not the case and it is possible to determine a subset of cells (a sort of "stripe" $\mathcal{S}$ from $(0, 0)$ to (n, m)) which contains P^* and whose cells can be evaluated without having to evaluate any cell out of $\mathcal{S}$. The total work to find P^* would then depend on $|\mathcal{S}|$ rather than being $\Theta(nm)$. The smaller $|\mathcal{S}|$, the better.

In this paper we outline an iterative procedure to determine such an $\mathcal{S}$, starting from a tentative small stripe and progressively increasing it until it can be proved that it is large enough: as shown in the results' section, the proposed approach outperforms DP if the strings are similar enough. We must underline that a recent result [1] shows that, under a strong conjecture similar to the $P \neq NP$ belief, no algorithm of worst-case complexity $O((nm)^t)$ with $t < 1$ is likely to exist for computing the edit distance. This however does not affect our result, which is to show that it is possible to have $t < 1$ in the *best*-case (while for Dynamic Programming best and worst case take the same time), or to have the same complexity as DP but with a better multiplicative constant.

2   Guesses and Stripes

If we reverse s' and s'', obtaining r' and r'', it is obvious that $d(s', s'') = d(r', r'')$. Assume $\mathrm{P}^r[\cdot, \cdot]$ is the DP matrix for $d(r', r'')$. Then, in the same way as $\mathrm{P}[i, j]$ represents the edit distance between a length-i prefix of s' and a length-j prefix of s'', we have that $\mathrm{P}^r[h, q]$ represents the edit distance between a length-h suffix of s' and a length-q suffix of s''.

To have a consistent indexing between the two matrices, let $\mathrm{S}[i, j] := \mathrm{P}^r[n - i, m - j]$. This way $\mathrm{P}[i, j] = d(s'[1, \ldots, i], s''[1, \ldots, j])$ and $\mathrm{S}[i, j]$ denotes the best cost of completing the transformation of s' into s'', turning the suffix $s'[i + 1, \ldots, n]$ into the suffix $s''[j + 1, \ldots, m]$. In particular, $\mathrm{P}[i, j] + \mathrm{S}[i, j]$ denotes *the optimal cost for turning s' into s'' given that the prefix $s'[1, \ldots, i]$ gets turned into $s''[1, \ldots, j]$.*

Our strategy will aim at calculating only a subset of cells of $\mathrm{P}[\cdot, \cdot]$ but, in order to do so, we will also need to calculate a subset of cells of $\mathrm{S}[\cdot, \cdot]$.

Let $w_0 := (0, 0)$ and $w_* := (n, m)$ be the upper left and the lower right cell, respectively. We say that two cells $a = (i, j)$ and $b = (u, v)$, with $a \neq b$, are *consecutive* (or *adjacent*) if $0 \leq u - i \leq 1$ and $0 \leq v - j \leq 1$. For two consecutive cells a and b we define a transition cost $\gamma(a, b)$ as follows: $\gamma((i, j), (i, j + 1)) :=$ $\mathtt{ins}(s''[j + 1]);$ $\gamma((i, j), (i + 1, j)) := \mathtt{del}(s'[i + 1]);$ and $\gamma((i, j), (i + 1, j + 1)) :=$ $\mathtt{sub}(s'[i + 1], s''[j + 1])$.

A *path* is a sequence $(v_0, \ldots, v_k)$ of cells such that $v_0 = w_0$, $v_k = w_*$ and, for each $t = 0, \ldots, k - 1$, v_t and v_{t+1} are consecutive. A path has length (or cost) $\sum_{t=0}^{k-1} \gamma(v_t, v_{t+1})$. Let OPT be the value of a shortest path, i.e., OPT $:= \mathrm{P}[w_*]$ (or equivalently, OPT $:= \mathrm{S}[w_0]$): we are seeking to determine OPT.

Our approach will require to make a sequence of guesses of the value of OPT, until we guess right. Given a certain guess $\tau \in \mathbb{R}$ let us define two sets of cells: $\mathcal{M}^0(\tau) := \{v : \mathrm{P}[v] \leq \tau/2\}$ (the top matrix part) and $\mathcal{M}^*(\tau) := \{v : \mathrm{S}[v] \leq \tau/2\}$ (the bottom matrix part).

Claim 1. *Let $\tau \in \mathbb{R}$. If $\tau \geq OPT$ then there exist consecutive cells $v \in \mathcal{M}^0(\tau)$ and $u \in \mathcal{M}^*(\tau)$ such that $\mathrm{P}[v] + \gamma(v, u) + \mathrm{S}[u] = OPT$.*

Proof. Let $P^* = (x_0, \ldots, x_t)$ be the optimal path, where $x_0 = w_0$ and $x_t = w_*$. If $x_t \in \mathcal{M}^0(\tau)$ then OPT $\leq \tau/2$. In this case, each $x_i \in \mathcal{M}^0(\tau)$ and also each $x_i \in$

$\mathcal{M}^*(\tau)$ so the claim is satisfied by taking $v = x_j$ and $u = x_{j+1}$ for any $j \leq t-1$. Otherwise, let j be the largest index $\leq t-1$ such that $x_j \in \mathcal{M}^0(\tau)$. Notice that this implies that $\mathrm{P}[x_{j+1}] > \tau/2$. Now, if it were $\mathrm{S}[x_{j+1}] > \tau/2$ we would have the contradiction $OPT = \mathrm{P}[x_{j+1}] + \mathrm{S}[x_{j+1}] > \tau \geq OPT$. Therefore, $x_{j+1} \in \mathcal{M}^*(\tau)$. By setting $v := x_j$ and $u := x_{j+1}$ we have $OPT = \mathrm{P}[v] + \gamma(v,u) + \mathrm{S}[u]$.

Given τ, we call *kissing pair* any pair of consecutive cells v and u such that $v \in \mathcal{M}^0(\tau)$ and $u \in \mathcal{M}^*(\tau)$. If kissing pairs exist, we say that $\mathcal{M}^0(\tau)$ and $\mathcal{M}^*(\tau)$ *kiss*, otherwise they are *apart*. Let $\mu(\tau)$ be the minimum value of $\mathrm{P}[v] + \gamma(v,u) + \mathrm{S}[u]$ over all kissing pairs (v,u). The following test gives a sufficient condition for a guess to be too small.

Claim 2. *Let $\tau \in \mathbb{R}$. If $\mathcal{M}^0(\tau)$ and $\mathcal{M}^*(\tau)$ are apart then $OPT > \tau$.*

Let r', c' be the largest row and column touched by $\mathcal{M}^0(\tau)$ and r'', c'' the smallest row and column touched by $\mathcal{M}^*(\tau)$.

Claim 3. *Let $\tau \in \mathbb{R}$. If $(r'' - r' > 1) \wedge (c'' - c' > 1)$ then $OPT > \tau$.*

The following lemma describes an optimality condition for a guess τ:

Lemma 1. *Let $\tau \in \mathbb{R}$. If $\mu(\tau) \leq \tau$, then $OPT = \mu(\tau)$.*

Proof. $\mu(\tau) \leq \tau$ implies that $\mathcal{M}^0(\tau)$ and $\mathcal{M}^*(\tau)$ kiss, then there is a path of length $\mu(\tau)$, so that $OPT \leq \mu(\tau)$. Assume $OPT < \mu(\tau)$. This implies $OPT < \tau$. By Claim 1, there is a kissing pair (v,u) such that $\mu(\tau) \leq \mathrm{P}[v] + \gamma(v,u) + \mathrm{S}[u] = OPT < \mu(\tau)$, which is a contradiction.

Given a set X of cells, let $\mathcal{S}(X)$ be the minimum set of consecutive cells, over the various rows, which contains all X. That is, if (i,a) is the first cell of X appearing in row i, and (i,b) is the last, then all cells $\{(i,a), (i,a+1), \ldots, (i,b)\}$ are in $\mathcal{S}(X)$ and this is true for all rows i containing elements of X. To build $\mathcal{M}^0(\tau)$ we need to compute $\mathrm{P}[v]$ for all $v \in \mathcal{S}(\mathcal{M}^0(\tau))$. We will show how to compute $\{\mathrm{P}[v] : v \in \mathcal{S}(\mathcal{M}^0(\tau))\}$ in time $O(|\mathcal{S}(\mathcal{M}^0(\tau))|)$: similar considerations hold for $\mathcal{M}^*(\tau)$. The sets $\mathcal{M}^0(\tau)$ and $\mathcal{M}^*(\tau)$ are similar to some diagonal "stripes" of cells: $\mathcal{M}^0(\tau)$ goes down diagonally from the upper-left corner while $\mathcal{M}^*(\tau)$ grows diagonally from the lower-right corner. Let us call $\mathrm{PART}(\tau) := \mathcal{M}^0(\tau) \cup \mathcal{M}^*(\tau)$ this partial DP matrix.

3 The Overall Procedure

For each guess τ, our procedure actually compute only the cells belonging to $\mathrm{PART}(\tau)$. Lemma 1 implies that we would like to make the smallest possible guess which triggers the condition $\mu(\tau) \leq \tau$. By Claim 1, we could use as guess an upper bound UB for OPT. Then, we would compute $\mathrm{PART}(\tau)$ (this can be done in time $O(|\mathrm{PART}(\tau)|)$) and find the kissing pairs (in the same time complexity), obtaining OPT. Unfortunately, $O(|\mathrm{PART}(\tau)|)$ is significantly smaller than $\Theta(nm)$ only if the upper bound is really tight (ideally, UB $\simeq$ OPT) and it should be computed extremely fast to be competitive with DP: no such quick and strong bound is known for the edit distance.

We therefore proceed "bottom-up", starting with a "small", optimistic guess LB $\leq OPT$ (a lower bound, or even $\tau = 0$) and then make a sequence of adjustments, increasing the guess until it is large enough to trigger the condition of Lemma 1:

1. Set $k := 0$, $\tau^0 := 0$ and compute PART(0)
2. **repeat**
3. $k := k + 1$
4.   Increase the guess: $\tau^k := \text{LB} + k\Delta$
5.   Compute PART(τ^k) from PART(τ^{k-1}) by left/right extending each strip
6.   Find the best kissing pair (v, u) in PART(τ^k)
7.   Set $\mu(\tau^k) := \text{P}[v] + \gamma(v, u) + \text{S}[u]$ ($\mu(\tau^k) := +\infty$ if there are no kissing pairs)
8. **until** $\mu(\tau^k) \leq \tau^k$

Step 1 is straightforward: PART(0) consists of two diagonals of 0s, one starting at w_0 and being as long as the longest common prefix of s' and s'', the other one starting at w_* and being as long as their longest common suffix.

Step 5 is incremental and needs some hints. We focus on updating $\mathcal{M}^0(\tau^{k-1})$ into $\mathcal{M}^0(\tau^k)$ (updating $\mathcal{M}^*$ is similar). We can assume, inductively on k, that for each row i we know the index $\alpha_{k-1}(i)$ of the first element such that $\text{P}[i, j] \leq \tau^{k-1}/2$ and the index $\omega_{k-1}(i)$ of the last element such that $\text{P}[i, j] \leq \tau^{k-1}/2$ (if there is no such element in row i, then $\alpha_{k-1}(i) := m + 1$).

In row 0 $\alpha_k(0) := \alpha_{k-1}(0) = 0$, so we only extend row 0 on its right. Starting at $j := \omega_{k-1}(0) + 1$ we compute all elements $\text{P}[0, j]$ and stop as soon as $\text{P}[0, j] > \tau^k/2$, setting $\omega_k(0) := j - 1$. Then, proceeding inductively on i, assume we have already extended the intervals at rows $0, \ldots, i - 1$ and are working on row i. We first extend the interval to the left of $\alpha_{k-1}(i)$. Notice that in the columns $0 \leq j < \alpha_k(i - 1)$ of row i there can be no entry of value $\leq \tau^k/2$, or there would have been also one in row $i - 1$, contradicting the definition of $\alpha_k(i - 1)$. Therefore, we start at $j := \alpha_k(i - 1)$ and compute the entries $\text{P}[i, j]$ from the adjacent entries $[i, j-1]$, $[i-1, j]$ and $[i-1, j-1]$ (clearly using only those whose value $\text{P}[]$ is known). We stop as soon as $j = \alpha_{k-1}(i)$ (or, if $\alpha_{k-1}(i) = m + 1$, we stop at $j = \omega_k(i - 1) + 1$). We set $\alpha_k(i)$ to be the first j found such that $\text{P}[i, j] \leq \tau^k/2$. Now we extend the strip to its right. Starting at $j := \omega_{k-1}(i) + 1$ we keep computing $\text{P}[i, j]$ from the known adjacent cells. We stop as soon as all the adjacent cells are not in $\mathcal{M}^0(\tau^k)$ and set $\omega_k(i) := j - 1$.

Notice how in step 6 the time needed to find the kissing pairs is bounded by $O(\min\{|\mathcal{M}^0(\tau^k)|, |\mathcal{M}^*(\tau^k)|\})$ rather than by $O(|\mathcal{M}^0(\tau^k)| \times |\mathcal{M}^*(\tau^k)|)$. Indeed, even if each pair (v, u) is made of an element $v \in \mathcal{M}^0(\tau^k)$ and another $u \in \mathcal{M}^*(\tau^k)$, to find the kissing pairs it is sufficient to scan all the elements of the smallest set, and, for each of them, look at the adjacent cells (3 at most) to see if they belong to the other set. This test can be done in $O(1)$ time. For example, to check if $(x, y) \in \mathcal{M}^0(\tau^k)$ we first check if $\alpha_k(x) \leq y \leq \omega_k(x)$. If that is the case, $\text{P}[x, y]$ is known and we check if it is $\leq \tau^k/2$.

Lower Bound and Guess Increment. The procedure we have outlined would work for any lower bound LB and any increment $\Delta > 0$ (indeed, as soon as τ^k becomes $\geq OPT$, by Claim 1 we find the optimal path), but its effectiveness depends on both these parameters. The best fine-tuning should be subject of further investigations, but we already found a quite good combination.

First we describe the lower bound used, which holds for all simple substitution matrices. Denote by s^L the longest between s' and s'' and by s^l the shortest. For each character $\sigma \in \Sigma$ and string $y \in \Sigma^*$, let $n_\sigma(y)$ be the number of occurrences of σ in y. For each symbol $\sigma \in \Sigma$ we define $\texttt{excess}(\sigma) = \max\{n_\sigma(s^l) - n_\sigma(s^L), 0\}$. Notice how the largest number of characters σ of s^l which could be possibly matched to identical characters in s^L is $n_\sigma(s^l) - \texttt{excess}(\sigma)$.

Claim 4. *The following is a valid lower bound to OPT, computed in time $O(n + m)$:*

$$LB = \left(\sum_\sigma \texttt{excess}(\sigma) \right) \times SUB + \left(|s^L| - |s^l| \right) \times IND$$

The proof that this is indeed a bound is omitted for space reasons. In our experiments we have noticed that this bound is quite strong when s' and s'' share a good deal of similarity.

In order to decide the step Δ with which we increase the guess, we opted to make this step proportional to the starting bound. By some tuning (not reported for space reasons) we determined that $\Delta := LB/3$ results overall in an effective procedure which terminates after a small number of iterations.

4   Computational Experiments and Conclusions

To assess the effectiveness of the proposed method we ran some experiments using an Intel® Core™ i7-7700 8CPU under Linux Ubuntu, equipped with 16 GB RAM at 3.6 GHz clock. The programs were implemented in C and compiled under gcc 5.4.0.

The problem instances were generated at random with a procedure based on two probabilities p_d and p_m, a base string length L, and the alphabet size. In each random instance, s' has length L and is randomly generated within the alphabet, while s'' is obtained by modifying s' as follows: each original character is deleted with probability p_d, then, if not deleted, it is mutated (randomly within the alphabet) with probability p_m. In the experiments reported here, we always used $p_d = p_m =: p$, leaving to further investigation the sensitivity to differing parameters. Clearly, lower values of p lead to more similar strings.

Table 1 reports the experiments run to compare the effectiveness of our method to the standard DP implementation. As it can be seen, the speedup is strongly dependant on p, i.e., on the string similarity: the more similar are the strings, the more effective our method is. However, even with $p = 0.2$ (i.e., the strings are quite dissimilar, differing approximately by 20% both in length and in contents), the proposed method saves about half time over DP, while it achieves an average speedup factor of 35× when $p = 0.01$.

Table 1. Comparison between our method and DP. Times are in seconds and the speedup is shown along with the filled percentage of the matrix. The alphabet size is 4.

String size		dna.mat				ones.mat				Average
		10000	20000	30000	40000	10000	20000	30000	40000	
$p = 0.01$	DP time	2.068	8.304	18.200	32.380	2.060	8.160	18.048	33.032	
	Our time	0.088	0.236	0.460	0.928	0.068	0.216	0.444	0.884	
	Speedup	23×	35×	40×	35×	30×	38×	41×	37×	**35×**
	Filled Perc.	1.8%	1.4%	1.2%	1.2%	1.3%	1.2%	1.1%	1.1%	1.3%
$p = 0.05$	DP time	2.020	8.048	18.480	32.484	2.080	8.220	19.032	32.480	
	Our time	0.352	1.164	2.848	5.944	0.236	0.868	2.040	4.076	
	Speedup	6×	7×	6×	5×	9×	9×	9×	8×	**8×**
	Filled Perc.	8.5%	7.4%	7.9%	8.3%	5.7%	5.1%	5.4%	5.7%	6.7%
$p = 0.1$	DP time	1.992	8.492	17.792	32.088	2.088	7.896	18.060	31.580	
	Our time	0.552	2.244	5.140	9.236	0.500	2.016	4.276	8.196	
	Speedup	4×	4×	3×	3×	4×	4×	4×	4×	**4×**
	Filled Perc.	13.9%	14.2%	13.5%	13.7%	12.3%	12.4%	11.9%	12.0%	13.0%
$p = 0.2$	DP time	1.772	7.548	16.092	28.524	1.780	7.104	15.900	32.976	
	Our time	1.140	4.528	9.912	18.864	0.996	3.908	8.260	15.456	
	Speedup	2×	2×	2×	2×	2×	2×	2×	2×	**2×**
	Filled Perc.	30.5%	29.1%	28.4%	28.9%	24.8%	23.8%	23.3%	23.6%	26.6%

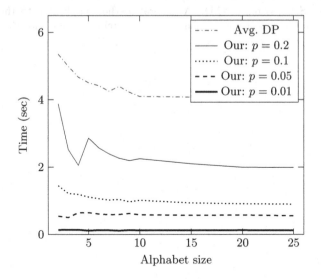

Fig. 1. Sensitivity of time performance towards the alphabet size as well as the mutation probability of strings, with base size equal to 15000 characters. DP is rather independent of p, thus we report an average time.

Another preliminary analysis, shown in Fig. 1, regards the sensitivity of the proposed method to the alphabet size. As can be seen, the performance are generally getting better with alphabets of increasing size, with a clear dependance on p.

5 Conclusions and Future Work

This preliminary extended abstract shows a promising approach to calculating the edit distance between two strings, leaving as future work the analysis of its performance with different lower bounds (or no lower bound at all), different τ increments as well writing the most effective code to readily identify and handle the kissing pairs.

References

1. Backurs, A., Indyk, P.: Edit distance cannot be computed in strongly subquadratic time (unless SETH is false). SIAM J. Comput. **47**(3), 1087–1097 (2018)
2. Gotoh, O.: An improved algorithm for matching biological sequences. J. Mol. Biol. **162**(3), 705–708 (1982)
3. Gusfield, D.: Algorithms on Strings, Trees, and Sequences: Computer Science and Computational Biology, 534 p. Cambridge University Press (1997)
4. Jones, N.C., Pevzner, P.A.: An Introduction to Bioinformatics Algorithms, 456 p. MIT Press (2004)
5. Levenshtein, V.I.: Binary codes capable of correcting deletions, insertions, and reversals. Doklady Akademii Nauk SSSR **163**(4), 845–848 (1965)
6. Needleman, S.B., Wunsch, C.D.: A general method applicable to the search for similarities in the amino acid sequence of two proteins. J. Mol. Biol. **48**(3), 443–453 (1970)

Efficient Transformation of Protein Sequence Databases to Columnar Index Schema

Roman Zoun[1(✉)], Kay Schallert[1], David Broneske[1], Ivayla Trifonova[1],
Xiao Chen[1], Robert Heyer[1], Dirk Benndorf[2], and Gunter Saake[1]

[1] University of Magdeburg, Magdeburg, Germany
{roman.zoun,kay.schallert,david.broneske,ivayla.trifonova,
xiao.chen,robert.heyer,gunter.saake}@ovgu.de
[2] Max Planck Institute for Dynamics of Complex Technical Systems,
Magdeburg, Germany
benndorf@mpi-magdeburg.mpg.de

Abstract. Mass spectrometry is used to sequence proteins and extract bio-markers of biological environments. These bio-markers can be used to diagnose thousands of diseases and optimize biological environments such as bio-gas plants. Indexing of the protein sequence data allows to streamline the experiments and speed up the analysis. In our work, we present a schema for distributed column-based database management systems using a column-oriented index to store sequence data. This leads to the problem, how to transform the protein sequence data from the standard format to the new schema. We analyze four different methods of transformation and evaluate those four different methods. The results show that our proposed extended radix tree has the best performance regarding memory consumption and calculation time. Hence, the radix tree is proved to be a suitable data structure for the transformation of protein sequences into the indexed schema.

Keywords: Trie · Radix tree · Storage system · Sequence data

1 Introduction

Mass Spectrometers are devices used to sequence proteins forming the building blocks of life [2]. By using mass spectrometry, bio-indicators of viruses, bacteria or species can be identified. Those are similar to a fingerprint and can help to diagnose diseases or optimize chemical processes, for example in bio-gas plants [3,7]. However, the measurement of a mass spectrometer takes up to two hours and is followed by a conversion of the measured data into a readable format. In order to gain insights into the converted data, an analysis step is required. Usually, this one is the protein identification [6]. The state-of-the-art

Supported by organization de.NBI and Bruker Daltonik GmbH.

protein identification approach uses a peptide-centric approach comparing the experimental data (mass spectra) with a protein sequence database. For this purpose, the proteins are divided into peptides, which results in billions of data sets. Each peptide is compared to every spectrum to find the highest similarity. As long as the measurement data is not written, the protein identification cannot start. This leads to a further delay of the protein identification. A method which identifies experimental data individually allows to analyze each single spectrum during the measurement as this one needs to be compared to all possible candidates [3]. Hence, an index schema is needed to query only suitable candidates of the sequence data and reduce the search area to a minimum [9,10].

In our work, we present the transformation process and describe four methods to aggregate the data into the index structure. The first one is the naive in-memory approach, the second one is the structured hard disk approach, the third method uses DBMS queries and the last one is the radix-trie-based method. At the end, we evaluate those methods and show that a trie structure is very efficient for the storage of sequence data and has the best overall performance among all approaches.

2   Background

In this Section, we provide some basic knowledge.

Protein Data: The involved data is the experimental data from the mass spectrometer device represented by a mass spectrum and the protein sequence data usually in FASTA format [1,3,4].

Real-Time Analytic of Mass Spectrometry Data: In an streaming environment, for each spectrum, we need to query only suitable candidates using a range query.

TRIE Data Structure: A radix tree extends the trie with the property that each parent node with only one child is merged with its child [5,8].

For our work, we want to use a radix tree with additional relation on each end node to a protein, because of the many-to-many relation between peptides and proteins.

3   Data Preparation for Real-Time Protein Identification

Indexed Masses of Peptides: To enable fast access to suitable candidates without losing all the information from the protein sequence database, we introduce our data structure. Our schema for the protein data consists of three tables – the protein table, the peptide table and the pepmass table. The `protein` table consists of minimum two columns – "UUID" and "Protein Sequence". For each uploaded protein sequence database a new column "Protein Data" is added to the table which contains the description from the FASTA file. The `peptide` table consists of minimum one column – "Peptide Sequence". Each uploaded protein sequence database appends new column – "Protein Set", to the `peptide` table which contains an unredundant collection of protein UUIDs. The `pepmass` table consists of "FASTA UUID", "Charge", "Pepmass" and "Peptide List" column.

Data Transformation: Our proposed schema increases the query performance to get the suitable candidates for each spectrum in a few milliseconds, which leads to increased storage because of the pre-calculations of all masses. As mentioned, we have to consider more than one protein sequence database. Hence a service is needed, which allows to upload new databases and to transform their data into our schema. In Fig. 1, we show the four steps to transform the data from the FASTA format into our schema.

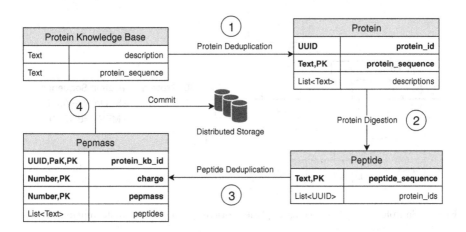

Fig. 1. Transformation steps from FASTA format to our indexed schema.

The first step is to deduplicate the protein sequences and merge the descriptions of the entries with a similar protein sequence. The next step is the protein digestion, which splits the protein sequence into smaller peptides. Equal peptide sequences can be extracted from different proteins, which leads to many-to-many relationship between proteins and peptides. Due to the high number of those relationships, the protein digestion is conducted with a list of protein id's in the table. The next step is the deduplication of the peptides during which only the unique peptides are left within the relation to the proteins that they come from. For each of those peptides the mass needs to be calculated with all the possible modifications and charges. Afterwards all the data is stored into the DBMS.

4   Implementation

In this section, we give a brief description of the implementation of the naive approaches and a detailed explanation of the extended radix tree approach.

The Transformation Using a Map Structure: The approach uses a map, a key value structure, for the peptide deduplication and the protein relationship.

The Transformation Using DBMS Queries: Since the result data needs to be inserted into the database management system (DBMS), we implemented queries to transform the data directly in the DBMS. We use UPDATE queries, which add new row if the key does not exist or otherwise, append a value to the list in the column.

The Transformation Using Extended Radix Tree Structure: In contrast, the ideas behind using a radix tree are more complex and, hence, we give a detailed explanation on the relational radix tree approach in the following.

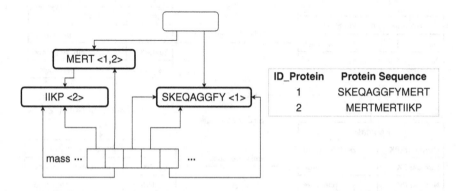

Fig. 2. Sample of a radix tree as peptide storage in the data transformation process.

In Fig. 2, we show the approach on an example. At the beginning a new column is added to the `protein` table and to the `peptide` table. For each protein, an identifier is generated and the protein descriptions in the protein table get updated. As next, the proteins get divided and generate a list of peptides for each protein. Hence, the two sample proteins "SKEQAGGFYMERT" and "MERT-MERTIIKP" are divided into peptides. The first protein generates the peptide list: "SKEQAGGFY", "MERT", and the second protein generates the peptide list: "MERT", "MERT", "MERTIIKP". Each of the peptides get inserted into the trie with the additional parameter of the protein identifier. However, the difference to the original radix tree is that we are not indexing the peptides, we link protein relationship to the peptides. To adapt the radix tree to our use case of peptide deduplication, we extend the usual radix tree, using a set of protein identifiers in each node. Hence, the trie contains at this moment all information for the **peptide** table. The link to the proteins is needed to identify the proteins, based on the peptide identification. The peptide sequences are automatically deduplicated.

The next step is to calculate masses for each peptide. The last step is to insert the data from the map with masses and the peptide sequences into the **pepmass** table.

5   Evaluation

For the evaluation, we use two different protein sequence databases - homosapiens and SwissProt.

Homosapiens is a small data set with 4794 proteins, 484,479 peptides and a storage size of 2.88 MB. SwissProt has the usual size of the productive used protein sequence databases with 255 MB storage size and 556,196 proteins and 37,403,696 peptides.

The goal of the evaluation is to find the best approach regarding the runtime and memory consumption.

Time Evaluation: For the time measurement, we measure the overall time of the approaches running the approaches 50 times.

`Homosapiens Data Set`: Regarding the homosapiens data set the naive in-memory approach is 5 s faster than the radix tree method. Furthermore, we can see that the SQLite and the DBMS approach are very slow and need hours of calculation time. The runtime is explainable because of the amount of writes on the slow hard disk. For the SwissProt Data Set, the naive approach needs 60 s for the homosapiens data set and 58 min for the SwissProt data set. The tree method takes 65 s for the smaller data set and 60 min for the SwissProt data set. The differences come from the calculation of the sequence recursively from the end node. In the trie structure, the mass is calculated traversing over the nodes while in the naive approach the whole sequence is accessible. This is not needed in the naive approach. The SQLite and the DBMS approach took days on the SwissProt data and is not comparable for such sizes of protein sequence databases.

Memory Consumption: In the memory consumption evaluation, we only consider the in-memory approach and the radix tree method. The other two methods failed due to their bad performance. For both data sets, the radix tree needs less memory to store all the peptide information with the relation to the proteins and their masses compared to the naive in-memory approach. The in-memory approach needs around one gigabyte for the homosapiens data set and over 100 gigabytes for the SwissProt data set. For the same data, a radix tree approach needs less then 300 megabytes for the homosapiens data set and around 14 gigabytes for the SwissProt data set. Moreover, the radix tree memory consumption increases between the data sets by factor 50, while the naive in-memory approach has a factor of 92. Overall, the radix tree is very beneficial for the peptide sequence data due to its inherent duplicate compression. Hence, the radix tree method combines in-memory speed and efficient data compression for sequence data.

Conclusion: In our work, we propose how to transform the protein sequence databases into this index schema. Therefore, we show the overall transformation and different implementation approaches of this transformation. After the evaluation of the four different methods, we conclude that an extended radix tree is the best structure for the peptide sequence data in order to transform the protein data into the index schema. The radix tree for peptide sequences combines nearly the best performance and minimal memory consumption. Hence, it proves to be beneficial for our use case of peptide deduplication.

Acknowledgments. The authors sincerely thank Niya Zoun, Gabriel Cam-pero Durand, Marcus Pinnecke, Sebastian Krieter, Sven Helmer, Sven Brehmer and Andreas Meister for their support and advice. This work is partly funded by the BMBF (Fkz: 031L0103), the European Regional Development Fund (no.: 11.000sz00.00.0 17 114347 0), the DFG (grant no.: SA 465/50-1), by the German Federal Ministry of Food and Agriculture (grants no.: 22404015) and dedicated to the memory of Mikhail Zoun.

References

1. Deutsch, E.W.: File formats commonly used in mass spectrometry proteomics. Mol. Cell. Proteomics **11**(12), 1612–1621 (2012)
2. Heyer, R., et al.: Metaproteomics of complex microbial communities in biogas plants. Microb. Technol. **8**, 749–763 (2015)
3. Heyer, R., et al.: Challenges and perspectives of metaproteomic data analysis. J. Biotechnol. **261**(Suppl. C), 24–36 (2017)
4. https://blast.ncbi.nlm.nih.gov/Blast.cgi?CMD=Web&PAGE_TYPE=BlastDocs& DOC_TYPE=BlastHelp:. Fasta format, November 2002
5. Leis, V., et al.: The adaptive radix tree: artful indexing for main-memory databases. In: IEEE International Conference on Data Engineering (ICDE 2013), pp. 38–49 (2013)
6. Millioni, R., et al.: Pros and cons of peptide isolectric focusing in shotgun proteomics. J. Chromatogr. A **1293**, 1–9 (2013)
7. Petriz, B.A., et al.: Metaproteomics as a complementary approach to gut microbiota in health and disease. Front. Chem. **5**, 4 (2017)
8. Shishibori, M., et al.: An efficient compression method for patricia tries. In: 1997 IEEE International Conference on Systems, Man, and Cybernetics. Computational Cybernetics and Simulation, vol. 1, pp. 415–420, October 1997
9. Zoun, R., et al.: Protein identification as a suitable application for fast data architecture. In: International Workshop on Biological Knowledge Discovery and Data Mining (BIOKDD-DEXA). IEEE, September 2018
10. Zoun, R., et al.: Msdatastream - connecting a bruker mass spectrometer to the internet. In: Datenbanksysteme für Business, Technologie und Web, March 2019

Cyber-Security and Functional Safety in Cyber-Physical Systems

Making (Implicit) Security Requirements Explicit for Cyber-Physical Systems: A Maritime Use Case Security Analysis

Tope Omitola$^{(\boxtimes)}$, Abdolbaghi Rezazadeh, and Michael Butler

Cyber-Physical Systems Research Group, Electronics and Computer Science,
University of Southampton, Southampton, UK
{tobo,ra3,mjb}@ecs.soton.ac.uk

Abstract. The increased connectivity of critical maritime infrastructure (CMI) systems to digital networks have raised concerns of their vulnerability to cyber attacks. As less emphasis has been placed, to-date, on ensuring security of cyber-physical maritime systems, mitigating these cyber attacks will require the design and engineering of secure maritime infrastructure systems. Systems theory has been shown to provide the foundation for a disciplined approach to engineering secure cyber-physical systems. In this paper, we use systems theory, and concepts adapted from safety analysis, to develop a systematic mechanism for analysing the security functionalities of assets' interactions in the maritime domain. We use the theory to guide us to discern the system's requirement, likely system losses, potential threats, and to construct system constraints needed to inhibit or mitigate these threats. Our analyses can be used as springboards to a set of principles to help enunciate the assumptions and system-level security requirements useful as the bases for systems' security validation and verification.

Keywords: Maritime security · Systems theory ·
System Theoretic Process Analysis (STPA) · Threat analysis ·
Cyber-Physical System Security

1 Introduction

With over 80% of global trade by volume and more than 70% of its value being carried on board ships and handled by seaports worldwide [1], the importance of a well-functioning national maritime industry cannot be over-emphasised. As key nodes in global transport chains providing access to markets, supporting supply chains, and linking consumers and producers, global ports are under constant pressure to adapt to changes in the economic, institutional, regulatory and

This work has been conducted within the ENABLE-S3 project that has received funding from the ECSEL Joint Undertaking under Grant Agreement no. 692455.

operating landscape. Many studies, e.g. [2], have identified maritime infrastructure and vessels to be potentially vulnerable to interference from cyber-threats. This potential vulnerability stems from a combination of increased connectivity and reliance on digital components, globally accessible navigation systems, and increasing levels of autonomous control. All such attacks have safety repercussions, with potentially serious impacts on human life, the environment and the economy. In this paper, we investigate how systems theory can be used for security requirements elicitation and analysis of an exemplar cyber-physical system (CPS), i.e., the communication systems enabling the interactions between maritime ships and their control centres.

2   Related Work on Security Analyses of Maritime Communication System

Due to their importance, safety and security impose constraining requirements that need to be fulfilled in the design and implementation of the communication systems of maritime assets. Maritime communication systems (MCS), as exemplar cyber-physical systems, have a coupling between the computational and physical elements, and correct system behaviour depends on correct functioning of the "interaction" of control logic with the physical system dynamics. Engineering such complex cyber-physical systems requires a holistic view on both product and process, where safety and security need to be incorporated across the engineering life-cycle to ensure such systems are safe from hazards and accidents, and secure from intentional and unintentional threats.

Traditional security analysis methods, such as **THROP** [6], work with threat models that are based on the fault-error-failure chain model. While these models are valid to describe threats to isolated components, they are insufficient to describe system threats in complex interconnected systems, as we have in modern CPS. **STRIDE** [7] takes a threat-centric approach to security analysis associating each threat with a particular asset from attackers' perspective. Although an advantage of STRIDE is that it helps change a designer's focus from the identification of specific attacks to focusing on the end results of possible attacks, one major disadvantage is that it mainly targets software systems.

In security, there can be a tendency to consider the assurance of security to be one of simply applying one particular solution, e.g. authentication or cryptography, or adhering to a best practice, such as threat modelling. But systems security, like safety and other system quality properties, is an emergent property of a system. This means that system security results from many things coming together to produce a state or condition that is free from asset loss, and the resulting loss consequences.

3   System Theoretic Process Analysis for Safety and Security Analysis

System Theoretic Process Analysis (STPA) [3] is an accident causality model based on systems theory, expanding the traditional model of causality beyond a chain of directly-related failure events or component failures to include more complex processes and unsafe interactions among system components. It is based on the three concepts of (a) safety constraints, (b) a hierarchical safety control structure, and (c) process models. STPA considers events leading to accidents occur because safety constraints were not successfully enforced.

STPA performs system safety hazard analyses, while our work focusses on system security. To some extent, system safety and security can be viewed as analogues of each other. Whilst system safety, and STPA in particular, focusses on analysing the system for potential accidents and identifying the hazards that could lead to those accidents, system security considers potential losses to the system and the associated threats that could lead to those losses, so that security constraints and mechanisms can be identified and integrated into the design to address the causes of these potential threats and to reduce the risk associated with the potential losses.

STPA has the following seven steps: (1) Stating system *purpose*, (2) Identifying *accidents*, (3) Identifying system *hazards* associated with accidents, (4) Constructing high-level *control structure*, (5) Translating system hazards into high-level *safety requirements*, (6) Identifying *unsafe control actions*, and (7) Using the results to *create or improve system design*.

Applying STPA concepts to security analysis, we have focussed on identifying **losses** (instead of accidents), **threats** (instead of hazards), translating the threats to a set of **security constraints** (instead of safety requirements), and identifying **insecure actions** (instead of unsafe control actions).

Figure 1 shows the entities of interest of our analysis, the MCS between an SBB controller and a Ship that the SBB controls. The first step in STPA is identifying the system's purpose.

3.1   System Purpose

Identifying the system purpose may require a few iterations, by the security analysis team, of what the system is supposed to achieve. After doing this, we identified the purpose of the MCS to be: "*the provision of timely, confidential, and correct communication of navigation data, acknowledgements and route updates, between SBB and Ship*". As our focus is on the MCS, we have made use of the two primitives of *SEND* and *RECEIVE* to model the actions of data being sent and received. The next stage is to identify the losses to the system. We start by defining what a loss is.

3.2   System Losses

A **loss** is a circumstance, event or operation that can adversely impact, and/or cause failure to, a system's purpose. We can see from Sect. 3.1 that unauthorised reading and modification of data, as well as any operation that can affect timely data reception will adversely affect our system's purpose. Taking these into consideration, the system losses, from the points of view of both the Ship and SBB, are listed in Table 1. From Table 1, we see the correspondence between a loss,

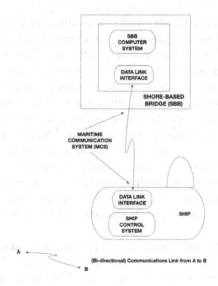

Fig. 1. Entities of interest (MCS, SBB, and Ship)

such as "Receiving Incorrect Ship Location data" (L_2), and its impact on one of the items of the purpose statement, i.e. "correctness". The next stage in STPA is to identify the system threats.

Table 1. System losses

Loss	Description of loss (from SBB's perspective)	Loss	Description of loss (from Ship's perspective)
L_1	Not receiving ship location data (at set periodic times)	L_5	Not receiving navigation data from SBB
L_2	Receiving incorrect Ship Location data	L_6	Receiving incorrect navigation data from SBB
L_3	Receiving Ship location/status data very late	L_7	Receiving navigation data very late from SBB
L_4	Un-authorised agent able to read Ship location/status data sent by Ship to SBB	L_8	Un-authorised agent able to read navigation data sent by SBB to Ship

3.3 System Threats

For the MCS to be effectively protected, we must resolve the security challenges inherent to that protection. These challenges can be looked at from the lens of the Confidentiality, Integrity and Availability, (C-I-A), triad. A loss of availability

Table 2. Possible system threats

Threat	Threat definition	Threat	Threat definition
T_1 Message congestion	Overload of the communication system persisting for a time significantly longer than service time	T_2 Interference	Unauthorised signal disruption
T_3 Tampering	Unauthorised data modification	T_4 Injection Attack	Introduction of false messages
T_5 Replay Attack	Valid communication is maliciously repeated or delayed	T_6 Relay Attack	Man-in-the-middle attack where all messages are forwarded verbatim between a valid sender and a valid receiver
T_7 Identity spoofing	Accessing a system disguised as a different actor	T_8 Loss of communications infrastructure	Unavailability of communication provisioning
T_9 Denial of service attack	Denial of service attack definition	T_{10} Traffic analysis	Unauthorised study of communication patterns between Ship and SBB
T_{11} Eavesdropping	Unauthorised listening to or reading of Ship and/or SBB's data and communication		

is the loss of the ability to access network resources. A loss of integrity is the intentional or unintentional changes to transmitted and stored data, while a loss of confidentiality is the unauthorised disclosure of transmitted and stored data.

A **threat** is a system state or a set of conditions that will lead to a system loss [3]. Table 2 shows the potential threats we identified for the MCS. These threats refer to specific opportunities by adversaries to defeat the system purpose and/or engender system losses. Table 3 shows the connection between threats and the losses they may cause, with a threat leading to more than one loss (e.g. T_1 being a causation factor to losses L_1, L_3, L_5 and L_7), or interactions of threats leading to a particular type of system loss (e.g. T_1, T_4, T_8, and T_9 as contributing factors to L_1).

3.4 System Control Structure

The control structure captures functional relationships and interactions of the main components of the MCS, as a set of command actions and feedback loops.

Table 3. System losses and threats

Losses & threats	L_1	L_2	L_3	L_4	L_5	L_6	L_7	L_8
T_1	X		X		X		X	
T_2		X	X	X		X	X	X
T_3		X				X		
T_4	X	X	X		X	X	X	
T_5		X		X		X		X
T_6				X				X
T_7				X				X
T_8	X		X		X		X	
T_9	X		X		X		X	
T_{10}				X				X
T_{11}				X				X

The questions to ask when constructing the control structure are: what are the main components in the system, what role does each play, and what are the command actions being used to interact. Figure 2 shows the control structures of the secure (Fig. 2A) and insecure (Fig. 2B) interactions between SBB and Ship.

3.5 Defining High-Level Security Constraints

After identifying the threats and constructing the control structures, the next major goal is to identify the security-related constraints necessary to prevent the threats from occurring. The question STPA enabled us to ask to help us identify the constraints was: "What constraints need to be in place to prevent the aforementioned threat conditions from occurring"? The constraints we identified are listed in (Table 4).

The security constraints together with the control structure helped us to answer questions such as: (a) what are we controlling (in the case of the MCS, it is data communication security between SBB and Ship), (b) what happens when the control actions go wrong (in our case, it is the likelihood that the identified threats may manifest), and (c) how can we mitigate those things we have identified can go wrong. Therefore, looking at the threats enumerated in Table 2, the security engineers can start identifying the constraints (Table 4) that need to be in place in order to mitigate those threats. For example, a threat such as "*an adversary interfering in the communication between SBB and Ship*" ($\mathbf{T}_2$), the system constraint is to guarantee against such threat occurring.

3.6 Identifying Insecure Actions

After the preliminary threat analyses carried out in Tables 2 and 4, the next step is to use STPA's four general categories of insecure control actions [3], to identify

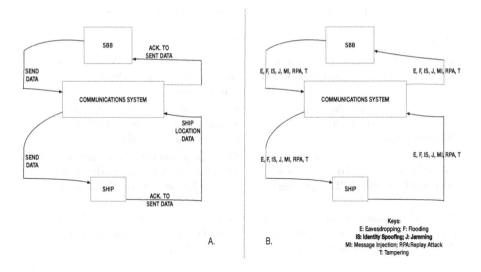

Fig. 2. Control structures for both the secure and insecure interactions between SBB and Ship

the conditions under which the actions of sending and receiving navigation data could lead to system threats. The security environment is dynamic, sometimes adversarial, with malicious actors that can learn how to subvert the system. In this kind of environment, the actions of interest to us are delineated as normal and malicious.

Malicious and Normal Control Actions. The system's normal control actions include: the *send*-ing and *receive*-ing of navigation and acknowledgement data from both the SBB and Ship. The malicious control actions are the actions a malicious agent is likely to make that can lead to a threat. These actions include: (a) Spoofing the address resolution and the IP protocols of the underlying communication network; (b) Eavesdropping actions; as well as (c) Traffic analysis actions.

Tables 5 and 6 present our use of STPA to analyse some controls and outputs issued by the MCS. **N.B.** We have used the following abbreviations in the two tables(**T**: Tampering, **I**: Interference, **IA**: Injection Attack, **RPA**: Replay Attack, **RLA**: Relay Attack, **IS**: Identity Spoofing, **DoS**: DoS Attack, **TA**: Traffic Analysis, **E**: Eavesdropping).

We found that the effect of the normal actions is binary: either they leave the system fool-proof or vulnerable to all of the threats described in Sect. 3.3, while the malicious actions expose the system to all of these security threats. This may be due to the existence of an operational communication system exposes it to such threats. In addition, our choice of primitives of *SEND* and *RECEIVE* are at a level that these states are binary, either a message is sent or not, and if a message is not sent, then no security threat arises.

Table 4. High-level security constraints for the MCS

Threat	System constraint
T_1 Communications requests exceed link capacity (**Message Congestion**)	SC_1 The system shall be able to prove the identity of agents during transactions
T_2 An adversary interferes in the communication between SBB and Ship (**Interference**)	SC_2 The system shall guarantee against communication interference between SBB and Ship
T_3 Valid communication between SBB and Ship is intercepted and data are maliciously modified (**Tampering**)	SC_3 The system shall maintain strong mutual continuous authentication, of SBB and Ship, during all operations' transactions
T_4 False messages, pretending to come from a valid source, are introduced into the system (**Injection Attack**)	SC_4 The system shall maintain strong mutual continuous authentication, of SBB and Ship, during all operations' transactions
T_5 The system maliciously repeats or delays valid communication between SBB and Ship (**Replay Attack**)	SC_5 The system shall maintain strong mutual continuous authentication, of SBB and Ship, during all operations' transactions
T_6 Valid communication is forwarded verbatim between SBB and Ship by a malicious agent (**Relay Attack**)	SC_6 The system shall maintain strong mutual continuous authentication, of SBB and Ship, during all operations' transactions
T_7 A malicious agent is pretending to be the SBB, the Ship or the Communication System (**Identity Spoofing**)	SC_7 The system shall maintain strong mutual continuous authentication, of agents, during all operations' transactions
T_8 There is discernible delay or a denial of service between the SBB and Ship (**Loss of Communications Infrastructure**)	SC_8 The system shall detect the loss of infrastructure
T_9 There is discernible delay or a denial of service between the SBB and Ship (**Denial of Service Attack**)	SC_9 The system shall ensure and maintain the specific turn-around time for each requested operation
T_{10} A malicious agent observes patterns of communication traffic between SBB and Ship (**Traffic Analysis**)	SC_{10} The system shall ensure protection over all communication
T_{11} An un-authorised agent listens into communication between SBB and Ship (**Eavesdropping**)	SC_{11} The system shall ensure that all the communications are not readable by any un-authorised party

Table 5. Normal control actions of the MCS

Normal control action	Not providing exposes system to threats	Providing exposes system to threats	Wrong time or wrong order exposes system to threats	Stopped too soon or applied too long exposes system to threats
SBB sends navigation data to ship	None	**UCA1**. T, I, IA, RPA, RLA, IS, DoS, TA, E	As in UCA1	As in UCA1
Ship receives Navigation data	None	As in UCA1	As in UCA1	As in UCA1
Ship sends Ack to SBB	None	As in UCA1	As in UCA1	As in UCA1
SBB receives Ack data from Ship	None	As in UCA1	As in UCA1	As in UCA1

Table 6. Malicious control actions of the MCS

Malicious control action	Not providing exposes the system to threats	Providing exposes the system to threats	Wrong time or wrong order exposes the system to threats	Stopped too soon or applied too long exposes the system to threats
Address resolution protocol (ARP) spoofing command	None	**UCA2**. IS, T, RPA, RLA, IA	As in UCA2	As in UCA2
IP spoofing command	None	As in UCA2	As in UCA2	As in UCA2
Packet tampering command	None	As in UCA2	As in UCA2	As in UCA2
Eavesdropping command (e.g. via a Network sniffer)	None	**UCA3**. Eavesdropping. Usually a passive attack	As in UCA3	As in UCA3
Traffic analysis command (e.g. via using Wireshark or P0f)	None	**UCA4**. Traffic Analysis. Normally a passive attack	As in UCA4	As in UCA4

3.7   Use Results to Create or Improve Design

The next step in STPA is to use the results of the analyses to help in designing a more secure system. Security is about risk management, and a purpose of risk management is to reduce losses. The C-I-A triad conjoined with the system losses and threats (Table 3) and with the high-level security requirements (Table 4) can help in assessing risk and weighting value. Depending on one's application domain, one can assign different values to the threats in Tables 3 and 4.

4   Conclusions and Future Work

This paper showed how systems theory and concepts from safety analyses, especially STPA, can be applied to the security analysis of a critical maritime infrastructure system (an exemplar cyber-physical system). We showed how STPA's systematic approach can help in eliciting the appropriate system's purpose, and to identify the losses and threats that may severely impact that purpose. We also showed how to derive the system constraints that can be used to inhibit or mitigate these threats, and described appropriate mitigation techniques.

For future work, we shall employ the Event-B [5] modelling methodology to help verify the completeness of the system constraints to mitigate or inhibit the threats that generated them.

References

1. United States Navy Biography. http://www.navy.mil/navydata/leadership/quotes. asp?q=253&c=6. Accessed 28 Nov 2018
2. UK Cabinet Office: National Cyber Security Strategy 2016 to 2021. UK Cabinet Office, November 2016
3. Leveson, N.G., Thomas, J.P.: STPA Handbook (2018)
4. Howard, G., Butler, M., Colley, J., Sassone, V.: Formal analysis of safety and security requirements of critical systems supported by an extended STPA methodology. In: 2nd Workshop on Safety & Security aSSurance (2017). https://doi.org/10.1109/EuroSPW.2017.68
5. Abrial, J.-R.: Modeling in Event-B: System and Software Engineering. Cambridge University Press, Cambridge (2010)
6. Dürrwang, J., Beckers, K., Kriesten, R.: A lightweight threat analysis approach intertwining safety and security for the automotive domain. In: Tonetta, S., Schoitsch, E., Bitsch, F. (eds.) SAFECOMP 2017. LNCS, vol. 10488, pp. 305–319. Springer, Cham (2017). https://doi.org/10.1007/978-3-319-66266-4_20
7. Potter, B.: Microsoft SDL threat modelling tool. In: Network Security, vol. 1, pp. 15–18 (2009)

Information Disclosure Detection in Cyber-Physical Systems

Fabian Berner$^{(\boxtimes)}$ and Johannes Sametinger

Department of Business Informatics, LIT Secure and Correct Systems Lab,
Johannes Kepler University Linz, Linz, Austria
`fabian.berner@outlook.de`

Abstract. The detection of information disclosure attacks, i.e. the unauthorized disclosure of sensitive data, is a dynamic research field. The disclosure of sensitive data can be detected by various static and dynamic security analysis methods. In the context of Android, dynamic taint-tracking systems like Taintdroid have turned out to be especially promising. Here we present a simulation environment, which is based on existing dynamic taint-tracking systems. It extends these and changes the analysis concept behind the taint-tracking system it uses. While taint-tracking is mainly used for mobile devices running Android, we postulate the importance of detecting information disclosure in any cyber-physical system. In this paper, we explore the detection of information disclosure by simulating devices and monitoring the information flows inside and among the devices.

Keywords: Information disclosure · Mobile computing ·
Taint-tracking

1 Introduction and Overview

Cyber-physical systems (CPSs) are smart systems including engineered interacting networks of physical and computational components [6]. Modern CPSs comprise systems of systems with heterogeneous components that face dynamic and uncertain environmental constraints [2]. Information disclosure is one potential danger for CPSs that can serve a direct purpose like *espionage*, *spamming* or provide *targeted advertising*. Stolen information can also be used by cyber criminals for other attacks like *social engineering*, *spoofing*, *phishing* or other *frauds*. A broad overview of Android security systems can be found for example in [9,11]. These and other similar studies discuss various security system approaches. Most papers that deal with novel taint-tracking systems also provide comparisons to distinguish their specific system from other ones, e.g., [10,12]. Our focus is on dynamic taint-tracking that can be used to detect information disclosure attacks by monitoring information flows between a data source and data sink.

The paper is organized as follows. In Sect. 2, we will evaluate taint-tracking systems, identify their shortcomings, and reveal possible future work in the context of CPSs. In Sect. 3, we introduce our simulation concept for the detection of

G. Anderst-Kotsis et al. (Eds.): DEXA 2019 Workshops, CCIS 1062, pp. 85–94, 2019.
https://doi.org/10.1007/978-3-030-27684-3_12

information disclosure, including a discussion on how simulations can be recorded and replayed. Section 4 concludes the paper.

2   Dynamic Taint-Tracking Analysis

Information disclosure attacks yield to the "unauthorized disclosure" of "sensitive data". In order to detect such attacks, we focus on *dynamic* taint-tracking techniques (or taint-analysis). In *dynamic* analysis systems, the object of investigation is executed and monitored, whereas a *static* analysis system analyses the source code without execution. Dynamic taint-tracking systems are used to track the flow of information from a data source to a data sink. The data to be monitored is marked in the data source through a kind of watermark called *taint*. The data sink can be used to retrace the data involved and whether a special protection requirement has been defined for the corresponding data source.

Taintdroid. *Taintdroid* by *William Enck*, the most known taint-tracking system, was published in 2010 by *Enck et al.* [5]. Taintdroid has been used by many security systems for Android, mostly in academia because of its advanced development and its availability as open source. Examples include *TreeDroid* [3], *VetDroid* [13], and *NDroid* [7].

TaintART and TainMan. *TaintART* was published in late 2016 [10]. Taintdroid is based on Dalvik Virtual Machine (VM), Dalvik was replaced by Android Runtime (ART) in Android 5.0 (2014). The prototypical implementation was done for Android 6.0 [10], which uses ART as runtime environment. The *TaintART compiler* integrates the taint logic into the compiled application. *TaintART runtime* executes the compiled native code and tracks tainted data. The public release of TaintART's source code has been announced but is not available yet as of this writing.

TaintMan is another unofficial successor of Taintdroid [12]. It integrates the taint-propagation logic in the apps and libraries to be analyzed.

Discussion. All mentioned taint-tracking systems share a subset of advantages and disadvantages, like limited system resources and complexity. For example, limited system resources are a problem, as security analysis at runtime on a mobile device costs valuable resources. Most of the security systems use a fully automated high-level analysis or a manual low-level analysis. A balanced medium between these two extreme analysis approaches is rare. A runtime security analysis either has to be fully automated or the user has to react manually to identify the potential security risks. Without expert knowledge in computer security, however, the average user may react to the security systems warning incorrectly [4]. Most taint-tracking systems can be used to detect information flows between apps. However, they are able to analyze only the information flows that use overt channels like Inter Process Communication (IPC).

Because of short release cycles software, we consider automated reruns of security analyses and subsequent comparisons of analysis results an important function of a security system. Our approach for such reruns will be presented in the next section.

3 Simulation Concept for Information Disclosure Detection

In order to overcome the shortcomings of existing systems, we have developed a concept for a simulation environment to evaluate security concerning the disclosure of information. As a general condition, we have considered it important not to change the properties and configuration of apps to be investigated or to modify them by code injection (e.g., in TaintMan). Apps with malicious functions can detect changes in the app properties or the internal program flow and then selectively deactivate the malicious functionality (evasion attack). Simulation enables test automation and automatic test replays, keeping test efforts for security officers or administrators as low as possible. First ideas of our simulation concept have been presented at [1].

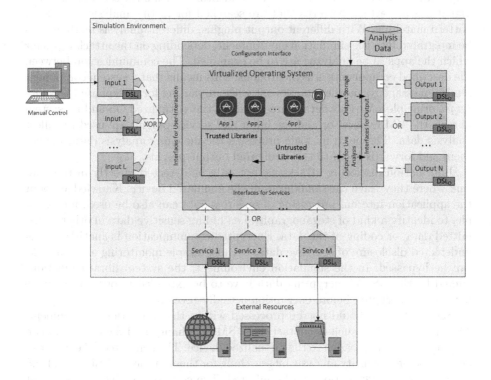

Fig. 1. Simulation environment

Figure 1 depicts an architecture overview of our system. The basis is a VM running a virtualized Android instance enhanced by an existing taint-tracking system. The virtualized Operating System (OS) executes the applications we want to evaluate. The taint-tracker is used to monitor the information flows between the applications and libraries on the virtualized device.

We consider three different plugin types for adapting the simulation environment: *Input plugins* can be used to simulate any user interaction. The user interaction can be done either by manual operation or by input generators like fuzzing tools. *Service plugins* can be used to simulate other mobile devices, other services, or other parts of the CPS. Other parts can either be fully simulated or be connected to the simulation environment via a stub service. The purpose of service plugins is to monitor communication with other devices. By simulating other devices, a security analysis does neither endanger the security of real devices, nor of any sensitive data and enables the detection of attack vectors that are distributed over several devices.

The security analysis itself, i.e., the evaluation of collected analysis data, can be implemented via *output plugins*. The analysis can be done either in parallel to the simulation execution (*live*) or after the simulation (*post-mortem*). Signature-based analysis methods are particularly suitable because mainly causal relationships and certain data patterns have to be identified within the security analysis. Signatures are attack patterns that are searched for in the analysis data using pattern matching. With different output plugins, different analysis methods can be integrated into the simulation environment, depending on the attack types for which the apps to be analyzed are to be examined. The communication between the described components is asynchronous. This means that a system component sends a message to another component without waiting for a response (*fire-and-forget* principle). Various data sources such as the virtualized OS and its system library, the VM hypervisor and the input and service plugins are used to collect analysis data. The program logic processes the incoming analysis data, merges them and forwards them to the configured output plugins.

Information disclosure attacks cannot only be detected directly in the data sink where they leave the boundaries of the simulated device. Knowledge about the application-internal processing of sensitive data can also be used, for example, to identify a kind of steganography, i.e., hiding sensitive data in other transmitted data, or coding of the data. Encrypted communication is another way of undetected disclosure of sensitive data, because simple monitoring at data sinks can be bypassed. In the simulation environment, the system library functions offered by the OS for encrypting data have to be extended in order to capture data to be encrypted before encryption actually takes place.

Sensitive data should not be processed within the simulation environment. Therefore, *fake data* such as contact data, SMS or changed GPS coordinates can be generated and uploaded to the virtualized device. The generated fake data can also be used for security analysis by searching for them in the analysis data, from which a flow of information can be inferred. The collected analysis data is sent to the output plugins and processed. The concept of our simulation environment does not provide for automatic reactions that prevent an information flow. Due to the generated fake data, no sensitive data is endangered during a simulation. Consequently, no automated reactions are necessary.

3.1   Simulation Process

A new simulation (heareafter *original simulation*) or a replay of an existing simulation (hereafter *replayed simulation*) can be started by configuring the simulation environment and plugins. The plugin selection and configuration depend on the goals of the analysis. The kind of user interaction for a new simulation can be defined by selecting the input plugin. For simulation replays, the user interaction is replayed automatically based on the recorded user interaction (*touch events*) and *system events* from the original simulation.

The selection of service plugins also allows different monitored communication partners for the simulation. However, the variation of service plugins for replayed simulations only makes sense when the variation of the simulated communication partners does not lead to an erroneous termination of the simulation. The output plugins for new simulations as well as for replayed simulation can be defined in advance of the security analysis. The analysis data is collected during the simulation itself.

Parallel to the simulation, the system and touch events from the virtualized OS are recorded for a later simulation replay. Depending on the selected output plugin, the security analysis is performed either in parallel or at the end of the simulation.

We can divide the simulation process into four main steps:

- Step 1: "Start and Configuration" The first step includes *start* and *configuration* of the simulation environment. We can either execute a simulation run or we can analyze collected data of a real device.
- Step 2: "Simulation" After initialization and configuration of the simulation environment, the actual simulation process can be started: the installed app can be started and, if necessary, preconfigured, e.g., a proxy server an app has to use).
  Depending on the chosen input and service plugins, simulation is executed automatically, half-automated or manually.
- Step 3: "Security Analysis" Step 3 includes the actual security analysis: the analysis data collected in the second step is evaluated. In case of a live analysis approach, this step is executed parallel to the simulation (second step). If post-mortem analysis is done, this step is started after the simulation has ended or, in case of a real mobile device, after the analysis data has been loaded.
- Step 4: "Evaluation and Reaction" It does not make sense to react to threats in a simulation environment. Nonetheless, in further development or of the output-plugins, an automated derive and recommendations to the user for action could be offered as an interesting new feature. These reactions can be tested by an automated simulation rerun after the reaction has taken place, e.g., restricted access to resources or repealed permissions.

The following reactions are suitable for most cases: The app can be released over a dedicated Secure Meta Market (SMM) when no unauthorized disclosure of sensitive data is detected. In case an unauthorized disclosure of sensitive data is detected, the app will either be rejected or provided in a dedicated SMM with additional security requirements. These security requirements could prescribe for example the usage of policy enforcement systems to impede the malicious function.

3.2 Security Analysis

A signature-based analysis approach is the most promising method when finding disclosed information in collected analysis data. Anomaly-based detection, instead, is suitable for detecting anomalies indicating possible malicious functionality. Most of the analysis data is in plain text and therefore easy to analyze by signature-based analysis.

Findings are indivisible, i.e., one finding refers to one information disclosure. Conversely, one information disclosure can be indicated by one or more findings. The simulation environment cannot distinguish between findings caused by legal or by illegal operations. *Legal* means that an operation, in which information leaves the secure context, is knowingly executed by the user. An *illegal* operation is either unknowingly executed by the user or based on an attack or malicious application. Illegal operations lead to unwanted information disclosure of sensitive data. We may not confuse the indivisibility of findings with *single-step attacks* and *multi-step attacks*. Single-step attacks can be detected based on one entry of the analysis data. Multi-step attacks can only be detected by analyzing two or more analysis data entries. Findings provide a logical unit and are therefore described as indivisible.

Regular expressions are an example for an easy-to-implement post-mortem approach for signature-based analysis. An example implementation of live analysis are off-the-shelf Complex Event Processing (CEP) systems. We propose the use of signature-based post-mortem analysis based on regular expressions, because they are fast, easy to use, and reliable. However, signature-based analysis is not able to detect zero-day exploits, signatures have to be developed and maintained, and data cannot be analyzed if it is encrypted, obfuscated or coded in some way. Alternative signature-based analysis techniques can be found for example in [8].

3.3 Recording Simulations

Our main idea is to collect as much input and output data as possible. Android distinquishes between touch and system events. *Touch events* comprise all user inputs and *system events* summarize all OS events such as incomming calls. To replay a simulation run, the recorded simulation data is processed to derive appropriate touch and system events that we can initiate. To later replay a simulation run, we first need to collect information about the system and its configuration including all the installed software. During the original simulation, any system and touch events have to be recorded with timestamps. Some communication data that cannot be analyzed by other analysis techniques can be collected at a library level, e.g., encrypted network connections. We adjust encryption methods provided by system libraries to log any encrypted data. Even if taint-tracking data, recorded network data and analysis data collected at service plugins are not replayed, they have to be recorded for later comparison of simulation runs. The security analysis results of a simulation run are valuable for a later simulation run comparison and therefore have to be collected, too. The collected analysis can be stored in files for a later replay.

3.4 Replaying Simulations

The ability to replay simulations helps users to work off the recurring activities faster. The idea is that the security analysis process does not have to be repeated manually, which would be time consuming and error-prone. Instead, we record the original simulation run and replay the simulation by repeating the system and touchevents automatically. Simulations can be replayed in a automated or half-automated way in case the simulation is initiated by the user. For an automated simulation replay, it would be possible to run the complete simulation environment as a cloud service. For full-automation, the simulation has to be started and executed without any manual user interaction. *Variation* as well as *degrees of freedom* pertain to the deviation of simulation replays. The difference between these two kinds of deviation is that variation describes an *unwanted* deviation of the simulation replay while degrees of freedom describe *wanted* derivations.

The question is how exactly a simulation can be replayed. Even if computers work strictly deterministic, there might be influence that is difficult to predict – especially in modern event-driven architectures. Accordingly, a simulation replay will variate in each replay. The purpose is to minimize the influence of variation on the simulation run and to avoid side-effects as far as possible. Beside the unwanted deviation that comes with variation, there are also degrees of freedom that can derivate a replayed simulation. Possible degrees of freedom define which wanted deviations are possible during a simulation replay. The usefulness of most degrees of freedom is self explanatory: they are useful since the release cycles of the apps and mobile OSs are very fast. By changing the app privileges and comparing the results of the original simulation and the replayed simulation run, possible reactions to mitigate the harm of malicious apps can be tested. By changing the app configuration (more apps, less apps or other apps installed), the simulation user can analyze if the malicious function depends on whether certain other apps are installed or not. Of course, this function can only be used to confirm or refute a certain suspect. By using this degree of freedom, two or more apps (from the same manufacturer), can be tested for the attack vector family *colluding apps*.

3.5   Comparing Simulations

To compare simulations, we have to define under which conditions two or more simulation runs are considered as equal. As discussed before, different degrees of freedom as well as variation can affect simulation replay. Thereby, the comparison also becomes more complex. To evaluate the equivalence of two or more simulation runs, the comparison should ideally detect causal relationships between incoming user input or rather system events, and the occurrence of information disclosure attacks. An example for a causal relationship would be when the user clicks a button and executes a malicious code which contains an information disclosure attack. We use different analysis data sources collected in different processes and threads. However, data collection can be more difficult because of scheduling and switching between the analysis data collection processes. Therefore, the detection of causal relationships between events and information disclosure attacks is impeded. We imagine a graphical simulation comparison tool in which the user can decide on the comparableness of two or more simulation runs. The idea is to represent simulation log entries, findings etc. in a diagram which can be visually analyzed by the user. For example, we can add supportive markings like coloring log entries and connection lines between equivalent nodes.

To support the user who wishes to evaluate the comparableness of different simulation runs, the comparison tool needs a feature to show and hide information the user does not need. Especially the amount of simulation log entries needs to be reduced, which can be done, by hiding corresponding entries from both compared simulation runs. Additionally, manual hiding of simulation log entries by marking entries could support the user to reduce the amount of simulation log entries to review. Even if findings occur at the same simulation sequence indicating the same information disclosure, the user cannot be sure that both findings lead back to the same attack or vulnerability. This happens due to the fact that we support black-box analysis of unknown applications. By evaluating the taint-tracking data, the finding can be backtracked to the code where it is initiated. Thereby, the findings become more comparable and the equality can be ascertained.

3.6   Sanity Checks

After a simulation replay, it is often unclear whether the simulation is replayed in a correct way and is thus comparable with the original simulation run. This can be due to variation and other deviations in the system configuration. On the other hand, the changed simulation run can be due to a failed simulation record or replay. Therefore, it makes sense to validate the simulation replay mechanism by a so-called *sanity check*. A sanity check can be useful to test the replay mechanism after a derivating simulation replay as well as when the simulation replay seems to be executed correctly. The procedure for sanity checks is as follows:

1. Simulation replay with the same configuration and software versions as in the original simulation run, which means replay of simulation without using any degree of freedom.
2. Recording of the simulation run.
3. Comparison of the findings and event order of both simulation runs.

The comparison of the findings in both simulation runs is not sufficient to test whether both simulations runs are equal. Even with the same findings, simulations can differ. For example, two totally different simulations may lead to the same findings, or simulation runs may differ in a simulation interval without findings, or findings may simply not be detected in both simulation runs.

If different findings for both simulation runs are found, one may not conclude that the simulation record and replay mechanism have failed. This can also be caused by a deviation outside the simulation context, for example, when a used web service is not available in one simulation run. Another possible drawback of sanity checks is that the influence of variation cannot determined and can thus distort the results of the sanity checks. All these possible problems have to be kept in mind during the sanity checks. Even if the results of sanity checks are in these exceptional cases no guarantee for the correct record and replay of simulation runs, they can be used in most cases to check whether the simulation was recorded and replayed correctly or not.

4   Conclusion

CPSs will increasingly handle sensitive information, thus information disclosure has to be prevented in order to avoid for example espionage. We have proposed a simulation environment that uses taint-tracking for the detection of such information disclosure. Simulation is useful as it allows to detect the revelation of information without endangering actual sensitive information. Additionally, it allows us to easily handle system updates, as simulation runs can be recorded, replayed and compared with previous runs. By simulating additional devices, we are able to detect information disclosure attacks, that are distributed over multiple devices. This makes it easy to find out subtle differences between different versions that may effect sensitive information. Taint-tracking is a mechanism used on the Android platform. We argue, that we need taint-tracking platforms for any system that handles sensitive information and that can be part of a CPS. In general, our concept can be ported and provided for other OSs as needed. We are currently developing the proposed simulation environment for the Android platform based on Taintroid. Experiments will be needed to fine-tune the recording, replaying and comparison mechanisms.

Acknowledgement. This work has partially been supported by the LIT Secure and Correct Systems Lab funded by the State of Upper Austria.

References

1. Berner, F.: Simulacron: Eine Simulationsumgebung zur automatischen Testwiederholung und Erkennung von Informationsabflüssen in Android-Applikationen. In: IT-Sicherheit als Voraussetzung für eine erfolgreiche Digitalisierung; Tagungsband ... 16. Deutschen IT-Sicherheitskongress, 21–23 May 2019, pp. 167–177 (2019)
2. Biro, M., Mashkoor, A., Sametinger, J., Seker, R.: Software safety and security risk mitigation in cyber-physical systems. IEEE Softw. **35**(1), 24–29 (2018)
3. Dam, M., Le Guernic, G., Lundblad, A.: TreeDroid: a tree automaton based approach to enforcing data processing policies. In: Proceedings of the 2012 ACM conference on Computer and communications security, CCS 2012, p. 894 (2012)
4. Enck, W.: Defending users against smartphone apps: techniques and future directions. In: Proceedings of 7th International Conference on Information Systems Security (2011)
5. Enck, W., et al.: TaintDroid: an information-flow tracking system for realtime privacy monitoring on Smartphones. In: Proceeding of the 9th USENIX Conference on Operating Systems Design and Implementation, OSDI 2010 (2010)
6. Cyber physical Systems Public Working Group: Framework for Cyber-physical Systems Release, 1, May 2016
7. Qian, C., Luo, X., Shao, Y., Chan, A.T.S.: On tracking information flows through JNI in android applications. In: 2014 44th Annual IEEE/IFIP International Conference on Dependable Systems and Networks (DSN), pp. 180–191. IEEE (2014)
8. Stallings, W., Brown, L., Bauer, M., Howard, M.: Computer Security: Principles and Practice. Always Learning, 2nd edn. Pearson, Boston and Mass (2012)
9. Sufatrio, Tan, D.J.J., Chua, T.W., Thing, V.L.: Securing android: a survey, taxonomy, and challenges. ACM Comput. Surv. **47**(4), 1–45 (2015)
10. Sun, M., Wei, T., Lui, J.C.S.: TaintART: a practical multi-level information-flow tracking system for android runtime. In: Katzenbeisser, S., Weippl, E. (eds.) Proceedings of the 2016 ACM SIGSAC Conference on Computer and Communications Security, pp. 331–342. ACM (2016)
11. Meng, X., et al.: Toward engineering a secure android ecosystem. ACM Comput. Surv. **49**(2), 1–47 (2016)
12. You, W., Liang, B., Shi, W., Wang, P., Zhang, X.: TaintMan: an art-compatible dynamic taint analysis framework on unmodified and non-rooted android devices. IEEE Trans. Dependable Secur. Comput. 1 (2017)
13. Zhang, Y., et al.: Vetting undesirable behaviors in android apps with permission use analysis. In: Sadeghi, A.-R., Gligor, V., Yung, M. (eds.) The 2013 ACM SIGSAC Conference, pp. 611–622 (2013)

Resilient Security of Medical Cyber-Physical Systems

Aakarsh Rao[1], Nadir Carreón[1], Roman Lysecky[1], Jerzy Rozenblit[1],
and Johannes Sametinger[2(✉)]

[1] University of Arizona, Tucson, AZ, USA
{aakarshrao7,nadir}@email.arizona.edu,
{rlysecky,jr}@ece.arizona.edu
[2] Johannes Kepler University Linz, Linz, Austria
johannes.sametinger@jku.at

Abstract. Incorporating network connectivity in cyber-physical systems (CPSs) leads to advances yielding better healthcare and quality of life for patients. However, such advances come with the risk of increased exposure to security vulnerabilities, threats, and attacks. Numerous vulnerabilities and potential attacks on these systems have been demonstrated. We posit that cyber-physical system software has to be designed and developed with security as a key consideration by enforcing fail-safe modes, ensuring critical functionality and risk management. In this paper, we propose operating modes, risk models, and runtime threat estimation for automatic switching to fail-safe modes when a security threat or vulnerability has been detected.

Keywords: Cyber-physical system · Medical device · Security

1 Introduction

Advancements in computational resources, sensors, and networking capabilities have led to the incorporation of Internet-connected devices in our lives [14]. These developments have also strongly influenced advances in healthcare and medical devices that have become part of digital health ecosystems [3]. Continual patient monitoring and services, interoperability, and real-time data access has become a normality. Life-critical devices, including implantable pacemakers and wearable insulin pumps, are essential for patients' health, well-being, and life. However, they pose additional security challenges in addition to those being considered for regular IT [17]. This is particularly exacerbated due to communication methods like Wi-Fi or Bluetooth that enable remote monitoring, real-time data analysis, and remote updates and configurations of device parameters [13].

2 Related Work

Considerable work has been done in the analysis of multi-modal CPSs with adaptive software for efficient resource utilization, incremental integration and

© Springer Nature Switzerland AG 2019
G. Anderst-Kotsis et al. (Eds.): DEXA 2019 Workshops, CCIS 1062, pp. 95–100, 2019.
https://doi.org/10.1007/978-3-030-27684-3_13

adaptability [8]. Mode change protocols are either event or time triggered [9]. Much work exists in real-time threat assessment and management particularly in intrusion detection systems, that however do not have the rigid timing and robust requirements found in medical CPSs [2,5]. Several works have been proposed for ensuring safety and security in medical devices, in broad areas of risk management, hardware devices, formal modeling and verification, and security schemes [4,15,19]. These proposed defenses require additional hardware to be worn by the patient or involve biological authentication schemes requiring further processing. However, security must be deeply integrated in the very design of medical devices and suitable mitigation schemes have to be incorporated in order to dynamically mitigate risks during deployment. Towards this direction we previously formally modeled a multi-modal software design framework with an adaptive risk assessment methodology and showcased preliminary findings with an integrated threat detector [11,12].

3   Resilient Security of Cyber-Physical Systems

Resilient context-aware medical device security has been proposed in [16]. The authors have shown the effect of sensitivity, impact, exposure, and authentication on context-awareness and resilience. We extend these mechanisms and propose to have resilience in CPSs by designing them in multiple modes, by modeling risk and by adaptive update of these risks, and eventually by automatic mitigation schemes.

3.1   Multi-modal Design

We propose to design application software for medical CPSs in a multi-modal fashion, cf. [8,12]. The system can operate in only one mode at a time. To ensure critical functionality of the medical device, the system has one essential mode that runs with a minimal attack surface. Each mode consists of a set of tasks to be performed by that mode, where a task would represent the implementation thread. In the essential mode, the tasks performed are the critical ones required for the essential functionality of the system. Different modes can have tasks in common based on the functionality.

3.2   Adaptive Risk Modeling

Risk modeling is a central activity in order to ensure security of systems [7]. A risk model is deeply integrated into the multi-modal software model by associating risk values at every hierarchical level of the mode to provide robust risk assessment and management, cf. [10]. During the deployment of the device, risks of the operations are assessed and updated based on the threats detected and estimated threat probabilities of the operations. Our threat detector is implemented in hardware and focuses on monitoring and analyzing the timing of the internal operations of the target system by utilizing a sliding window [6]. At runtime, the timing samples inside each sliding window are analyzed, and the probability of the current execution being malicious (threat probability) is calculated. In addition to the proposed risk update in [11], we augment an additional risk update condition for impactful operations. Impactful operations are defined as operations whose base risk is beyond an impact threshold that would directly affect the critical functionality of the system. The risk update is exponential for these operations as compared to an additive increase as proposed in [11].

3.3 Threat Estimation

For runtime risk assessment, risk values need to be updated in a composite risk model. If we detect security threats with estimated threat probabilities, we can update risk values accordingly, depending on the estimated security threat probabilities. We assign initial composite risks to the modes based on their composition of tasks or task options that constitute the modes. For example, initial operation risks can be assigned based on security scores as proposed in [18].

3.4 Automatic Mitigation Schemes

In many domains, including health, risks have arisen through the addition of software and connectivity. Attack vectors that did not previously exist have suddenly become a priority [1]. To ensure risk management during deployment of the device, we propose an automatic mitigation scheme that changes operating modes of the system triggered by updates in risk values in order to reduce the effective risk of the system. The system risk is the risk of the current operating mode. A system level risk threshold is defined by an expert that represents the level beyond which the system cannot operate in the current operating mode. It is assumed that during initial deployment the system always operates in the highest mode, thus, having full funtionality and connectivity to the outside world.

3.5 Architectural Overview

Figure 1 gives an overview of the components of a secured cyber-physical system. We can see that various modes are available that are switched depending on risk assessment. Depending on the determination of risks, threat estimations will lead to mitigation activities that have an effect on the operation of the CPS by means of switching the modes. The modes have common functionalities, but the lower the mode number, the more restrictive are activities that may lead to security problems. We imagine that in the essential Mode 0, a CPS will only provide basic functionlity with any communication turned off that is not absolutely necessary for the basic functioning. Thus, Mode 0 will have the smallest attack surface possible, while Mode n will provide full functionality of the system with the biggest attack surface.

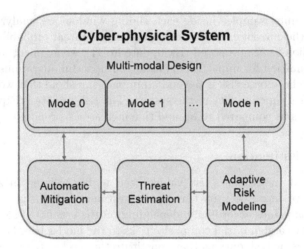

Fig. 1. Architectural overview.

4 Insulin Pump

We have started to evaluate our proposal with different insulin pump scenarios. The attacks we will use in these scenarios will be based on known malware that we will adapt to the insulin pump model. For example, the Fuzz malware is a common attack by malicious users, with the purpose of interfering with the predefined functionality of the target system by "fuzzing", or slightly altering the data. The Information Leakage malware is another well-known attack, with the goal of breaking confidentiality by extracting information from the patient and transmitting it to an unauthorized user. We plan to use the same configuration for all simulations. The starting point of the simulations will always be the highest functionality mode. Then we'll try to find out how well the system will adapt to different threat scenarios and whether these adaptations will effectively be able to mitigate the threats.

5 Conclusion

CPSs pose many security threats. We suggest that, in addition to considering security issues during development from the very beginning, we have to make sure that our systems are capable of reacting to threat scenarios not yet known during development. Software updates are a means of adapting systems in such scenarios. However, for CPSs, updates and patches are not always practicable. For such cases, our proposed resilience mechanisms with a multi-mode design, adaptive risk updating, and an automatic mitigation scheme seems a reasonable solution. We are now in the process of experimenting with an insulin pump to find out how our proposed solution reacts to various attack scenarios.

Acknowledgement. This work has partially been supported by the LIT Secure and Correct Systems Lab funded by the State of Upper Austria.

References

1. Biro, M., Mashkoor, A., Sametinger, J., Seker, R. (eds.) Software safety and security risk mitigation in cyber-physical systems. IEEE Softw. **35**(1), 24–29 (2018)
2. Blyth, A., Thomas, P.: Performing real-time threat assessment of security incidents using data fusion of IDS logs. J. Comput. Secur. **14**(6), 513–534 (2006)
3. Krishnamurthy, R., Sastry, A., Balakrishnan, B.: How the internet of things is transforming medical devices. Cognizant 20–20 Insights, Cognizant (2016)
4. Li, C., Raghunathan, A., Jha, N.K.: Improving the trustworthiness of medical device software with formal verification methods. IEEE Embed. Syst. Lett. **5**, 50–53 (2013)
5. Lu, S., Seo, M., Lysecky, R.: Timing-based anomaly detection in embedded systems. In: Proceedings of the 20th Asia and South Pacific Design Automation Conference, pp. 809–814 (2015)
6. Lu, S., Lysecky, R.: Time and sequence integrated runtime anomaly detection for embedded systems. ACM Trans. Embed. Comput. Syst. **17**(2), 38:1–38:27 (2018)
7. National Institute of Standards and Technology: Guide for Conducting Risk Assessments. NIST Special Publication 800–30 Revision 1, September 2012
8. Phan, L.T.X., Lee, I.: Towards a compositional multi-modal framework for adaptive cyber-physical systems. In: IEEE International Conference on Embedded and Real-Time Computing Systems and Applications, pp. 67–73 (2011)
9. Phan, L.T.X., Chakraborty, S., Lee, I.: Timing analysis of mixed time/event-triggered multi-mode systems. In: IEEE Real-Time Systems Symposium (RTSS), pp. 271–280 (2009)
10. Rao, A., Rozenblit, J., Lysecky, R., Sametinger, J.: Composite risk modeling for automated threat mitigation in medical devices. In: Proceedings of the Modeling and Simulation in Medicine Symposium, Virginia Beach, VA, USA, pp. 899–908 (2017)
11. Rao, A., Carreon Rascon, N., Lysecky, R., Rozenblit, J.W.: Probabilistic security threat detection for risk management in cyber-physical medical systems. IEEE Softw. **35**(1), 38–43 (2018)
12. Rao, A., Rozenblit, J., Lysecky, R., Sametinger, J.: Trustworthy multi-modal framework for life-critical systems security. In: Annual Simulation Symposium, article no. 17, pp. 1–9 (2018)
13. Roberts, P.: Intel: New Approach Needed to Secure Connected Health Devices (2015). https://www.securityledger.com/2015/03/intel-new-approach-needed-to-secure-connected-health-devices/
14. Rose, K., Eldridge, S., Chapin, L.: The Internet of Things (IoT): An Overview-Understanding the Issues and Challenges of a More Connected World. Internet Society (2015)
15. Rostami, M., Juels, A., Koushanfar, F.: Heart-to-Heart (H2H): authentication for implanted medical devices. In: ACM SIGSAC Conference on Computer & Communications Security, pp. 1099–1112 (2013)
16. Sametinger, J., Steinwender, C.: Resilient context-aware medical device security. In: International Conference on Computational Science and Computational Intelligence, Symposium on Health Informatics and Medical Systems (CSCI-ISHI), Las Vegas, NV, USA, pp. 1775–1778 (2017)

17. Sametinger, J., Rozenblit, J., Lysecky, R., Ott, P.: Security challenges for medical devices. Commun. ACM **58**(4), 74–82 (2015)
18. Sametinger, J., Rozenblit, J.W.: Security scores for medical devices. In: Proceedings of the 9th International Joint Conference on Biomedical Engineering Systems and Technologies (BIOSTEC 2016) - Volume 5: HEALTHINF, pp. 533–541 (2016)
19. Xu, F., Qin, Z., Tan, C.C., Wang, B., Li, Q.: IMDGuard: securing implantable medical devices with the external wearable guardian. In: IEEE INFOCOM (2011)

Securing Smart Homes

Johannes Sametinger[(✉)]

Department of Business Informatics, LIT Secure and Correct Systems Lab,
Johannes Kepler University Linz, Linz, Austria
`Johannes.Sametinger@jku.at`

Abstract. Devices in people's homes increasingly depend on software
and hardware components. They interoperate with other devices wire-
lessly and through the Internet. The sensitive nature of some of their
data, their increasing interoperability puts their security at the fore-
front. In this paper we will show smart homes can resiliently be secured
by varying functionality, exposure and authentication. We will demonstrate these mechanisms on simple smart home situations.

Keywords: Security · Smart home · Functionality · Exposure ·
Authentication · Resilience

1 Introduction

The Internet of Things (IoT) is touching many aspects of our lifes. Smart homes
are just one such example. The smart home idea is that all gadgets and devices
will be connected online and can be controlled anywhere over the Internet, such
as on our smartphones [2]. In many domains, including smart homes, risks have
arisen through the addition of software and connectivity. Attack vectors that did
not previously exist have suddenly become a priority [3]. Connected devices of a
smart home include light bulbs, light switches, security cameras, TVs, door locks,
smoke alarms, thermostats, ceiling fans, etc. All these devices can be controlled
by their user(s), e.g., via smarthphone, or they control each other in some way,
e.g., the WiFi-connected ceiling fan may get controlled by a thermostat. Smart
homes result in convenience and peace-of-mind in many situations. Answers to
the following questions are only a few taps and wipes away. Did the baby-sitter
get locked out of the house? Did we leave the lights on? Did we forget to lower
the thermostat? Additionally, GPS-enabled pet trackers report our pet's location and the temperature there. Smart homes comprise more than just remote
controls, they should anticipate our needs, keep us healthy and save us money.
Thermostats can learn our preferred settings and schedule, lights turn on and off
as we come and go, and refrigerators may adjust their temperatures according
to how much food they hold [14].

The number of connected devices generates an important discussion point:
security. Connected devices give hackers potential opportunities to break into
homes, e.g., monitoring cameras, smart TVs or even connected baby monitors.

© Springer Nature Switzerland AG 2019
G. Anderst-Kotsis et al. (Eds.): DEXA 2019 Workshops, CCIS 1062, pp. 101–107, 2019.
https://doi.org/10.1007/978-3-030-27684-3_14

Security is about protecting information and information systems from unautho-rized access and use. Confidentiality, integrity, and availability of information are core design and operational goals. Software security is "the idea of engineering software so that it continues to function correctly under malicious attack" [7]. In this sense, the goal of smart home security is to engineer smart devices in people's homes so that they ideally would be immune to malware implantation or if a breach occurred, they would continue to function correctly. One form of protection is the use of firewalls as proposed in [15]. A form of resilience in the context of smart homes has been proposed in [5]. Doan et al. ask themselves what if the cloud is not available in a setup where the cloud is used for analytics, computation and control. Disconnection from the cloud can happen not only by accident or natural causes, but also due to targeted attacks. Our reflections are independent from whether a cloud is used. We will base our work on security scores presented in [16]. Additionally, we will adapt an approach to resiliently secure medical devices that has been proposed in [17].

This paper is structured as follows: In Sect. 2, we provide an introduction to smart homes. Security scores will get introduced in Sect. 3. Securing regular IT is the topic of Sect. 4. Section 5 is about securing smart home devices. A conclusion follows in Sect. 6.

2   Smart Homes

The term smart home is commonly used to define a home with many smart devices. They include appliances, lighting, heating, air conditioning, TVs, com-puters, entertainment systems, security, and cameras. The devices are capable of communicating with one another. Users can control them via time schedule, from any room in the home, as well as remotely from anywhere in the world, typically via smartphone. Smart homes can be categorized into home appliances, lighting and climate control systems, home entertainment systems, home com-munication systems, and home security systems [11]. Smart home technology has evolved from basic convenience functionality like automatically controlled lights and door openers to benefits like water flow sensors and smart meters for energy efficiency. IP-enabled cameras, motion sensors, and connected door locks offer better control of home security [8].

Attackers can manipulate smart devices to cause physical, financial, and psy-chological harm. Examples include hidden access codes in smart door locks or a fire at a victim's home caused by a targeted smart oven [4]. Successful hacking of smart home devices has been demonstrated on several occasions. For exam-ple, hackers used smart home devices, such as CCTV cameras and printers, to attack popular websites [1]. Twitter, Spotify, and Reddit were among the sites taken offline. A description of several plausible attacks on smart home systems using software-defined radio, in particular using the communications protocol ZigBee, is given in [6]. Design flaws of a smart home programming platform were exploited by researchers to construct proof-of-concept attacks to secretly plant door lock codes, to steal existing door lock codes, to disable vacation mode of a home, and to induce fake fire alarms [8]. Threats of eavesdropping and direct compromise of various smart home devices were shown in [4]. A garage system that can be turned into a surveillance tool that in turn informs burglars when the house is possibly empty. A controller's microphone that can be switched on to eavesdrop on conversations. A controller that has authentication bypass bugs and cross-site request forgery flaws [9]. The German computer magazine c't describes how a security vulnerability could have led to burglars entering homes with the press of a button. Criminals could have had access to a victim's home without leaving any traces with only a single tap on their smartphone, a nightmare for home-owners and tenants, as well as insurance companies [10].

3   Scoring Smart Home Devices

The importance of security measures for medical devices has been scored based on whether they process or communicate sensitive information, whether they process or communicate safety-critical information, and how exposed they are to their environment [16]. These characteristics give an indication of the extent to which security countermeasures have to be taken upon device development. Additionally, the existence of vulnerabilities of a device gives an indication of how urgent there is a call for action when a device is in use. The scores introduced in [16] can also be used for non-medical devices.

Sensitivity. Sensitive information includes anything about an individual, e.g., name, address, current location, location history. A device's sensitivity level indicates the amount of sensitive information on that device. A sensitivity score provides an estimate of the amount of sensitive information on a device, and an estimate of how easily we can attribute information to a specific person.

Impact. Devices differ in the degree of impact that they can have on its environment. Typically, devices with a high benefit (utility) also pose a high potential harm. For example, a smart garage door may open whenever a specific smartphone comes close enough, making a home owner's coming home more convenient by not having to open the door manually. The door can pose a threat if it malfunctions, because thieves may be able to open the door when the owner is away. The impact of a device indicates the potential benefit and the potential harm a device can do, be it directly or indirectly. Theft is an indirect impact of a hacked garage door. A direct impact may involve ruined motors when someone foolishly plays around with them.

Exposure. High interoperability means that a device is highly exposed to its environment. Interoperability refers to the mode in which devices work with one another. Devices can operate as stand-alone (low exposure) or they can interoperate with other devices via a connection to the Internet (high exposure). High exposure offers a big attack surface for potential intrusions. However, it also provides flexibility and benefits. For example, a home owner may check the temperature in his house and make adjustments before coming home from a trip.

Vulnerability. We have stated in the introduction that secure devices have to continue to function correctly even if under a malicious attack. Vulnerabilities are errors in devices, typically in software, which attackers can directly use to gain unauthorized access to the device. They pose a threat to the device itself, to the information it contains, to other devices that communicate with it, and to the environment in which it operates. When we detect vulnerabilities, the threat to a device can increase rapidly. It is important for everyone involved to have a clear picture of the current security status and to make reasoned decisions about needed steps in order to decrease the threat, if needed.

Privacy and Safety Risks. A device is at risk when it is vulnerable and stores sensitive information, or when it is vulnerable and it has a physical impact on its environment. Privacy concerns exist when personally identifiable information or other sensitive information is processed or stored. Insufficient access control is typically the root cause of privacy issues. Disclosure of sensitive information may result in negative consequences. For example, potential thieves may get the information that nobody is at home.

4 IT Security

Securing an IT infrastructure has many facets, ranging from securing networks, cf., firewalls, intrusion detection, to securing computers, cf., antivirus software, and to securing code, cf. security bugs and flaws. We need to protect information and information systems from unauthorized access and use. The core goals are confidentiality, integrity and availability of information.

Weaknesses and Vulnerabilities. Software weaknesses are "flaws, faults, bugs, vulnerabilities, and other errors in software implementation, code, design, or architecture that, if left unaddressed, could result in systems and networks being vulnerable to attack. Example software weaknesses include: buffer overflows, format strings, etc." [12]. An information security vulnerability is "a mistake in software that can be directly used by a hacker to gain access to a system or network" [13]. While a weakness describes a software bug in general, a vulnerability is a specific instance of such a weakness. For example, CVE-2019-7729 describes an issue of a smart camera where due to setting of insecure permissions, a malicious app could potentially succeed in retrieving video clips or still images that have been cached for clip sharing. Prominent software security bugs include buffer overflows, SQL injections and crosssite scripting. There are many recent examples, where these bugs have occurred and caused damage. While security bugs are problems at the implementation level, security flaws are located at the architecture or design level. Security flaws are much harder to detect and typically need detailed expert knowledge.

Updates and Patches. Software vulnerabilities are often the entrance door for attackers, which software vendors have to fix in their updates. Microsoft and Apple, for example, periodically offer updates to their operating systems, including many security fixes. Newer versions of their operating systems automatically download and install such updates. Users are strongly encouraged to keep their systems up-to-date. Security fixes typically address specific entries in the CVE database, e.g., a new software version of a home security system may resolve the vulnerability described in CVE-2019-7729. If users refrain from updating to the newer version, they are vulnerable to attackers who have public access to a description of this particular vulnerability.

Risk Reduction. We can see privacy and safety risks as a combination of vulnerability and sensitivity as well as impact. Securing devices means reducing these risks. We reduce risks by eliminating vulnerabilities. We could also reduce sensitivity or impact. For example, if we eliminate sensitive information from a server, then we are less prone to leaking sensitive information, no matter what the vulnerability of the server. The same is true for impact. If we have a way of disabling specific functions of an airplane control system, an attacker will have fewer opportunities to cause havoc, unless she can activate the functions again by software. Typically, we do not sacrifice functionality for the sake of security. We normally eliminate known vulnerabilities by developers and patch used systems with new improved versions.

5   Smart Home Security

It is not always possible to update and patch smart home devices like regular computers. These devices often do not provide an automatic update mechanism. That per se would additionally increase the device's attack surface. We argue that if it is not an option to update software in order to eliminate vulnerabilities, we alternatively can adjust sensitivity, impact and exposure. Additionally, we can enforce stricter authentication measures in order to reduce security risks.

We will use a smart garage door opener and smart video door bell to illustrate our ideas. Users can monitor and control their garage door from anywhere using their smartphone or tablet. Many smart garage door openers connect directly to a home owner's Wi-Fi network or via a Bluetooth connection. They can be set up in minutes using a smartphone.

There is always a tradeoff between security and usability. *IEEE Security & Privacy* has published a special issue on this topic [18]. While we may give usability a higher priority in situations where we regularly use, say, a garage door, we may volunteer for higher security in situations where we may be on vacation. We may also want to have stricter authentication to surveillance cameras while we are living at home, but less strict authentication when on vacation, to allow us easier access to surveilling our home. Current systems do not support such a tradeoff between usability and security.

Functionality. A garage door opener has quite a simple functionality. It can open and close a garage door. And it allows to do so remotely. This is a convenient feature when we commute to work on every weekday. The associated risk with a garage door opener is that someone may get access that is not supposed to have access, e.g., a thieve. Someone may also get sensitive information, e.g., a schedule of when we typically leave home in the morning and return in the evening. This information is useful even when the opener will not be used to enter the house.

Should we be on vacation, we may want the garage door opener to be turned off, providing access to no one. We may also want the garage door opener to log opening and closing the door in regular intervals, just in case someone is reading that information without authorization.

Exposure. If the smart video door bell is connected to the Internet, we can have a look at who's in front of the door remotely. We can also open the door remotely, for example, to let someone in to deliver a package. If our smart home gets hacked, then someone else may do the same. Suppose, we go on a three week vacation, we may appreciate the fact that we can let someone trusted into the house. But we may also feel better to know that nobody will get in, not even by hacking. If we refrain from being able to let someone in by disconnecting from the Internet, we also keep others out.

Authentication. Rather than completely refraining from getting remote access to either the garage door opener and the smart video door bell, we may also choose a path in the middle by strengthening authentication. Thus, we may provide remote access, but only with additional authentication. This may be impractical for everyday use, but may be a good compromise when we want our home to be better protected, but when we still want to have access if needed.

6 Conclusion

We argue that the trade-off between usability and security can be challenging with smart homes. Having everything connected to the Internet is useful in many situations. But it increases the attack surface and makes keeping the home secure more challenging. Of course, we have to make sure not to use default passwords and to keep systems up-to-date. But sometimes, it may feel better being able to disconnect and later connect again. Disconnecting may be an option if we learn about vulnerabilities and exploits of a device that we are using. We may later decide to connect again when these vulnerabilities have been fixed. Connecting again may even be done remotely by using strict authentication.

Acknowledgement. This work has partially been supported by the LIT Secure and Correct Systems Lab funded by the State of Upper Austria.

References

1. BBC: 'Smart' home devices used as weapons in website attack, BBC Technology, 22 October 2016. http://www.bbc.co.uk/news/technology-37738823
2. Bilton, N.: Pitfalls of the connected home. The New York Times, 16 October 2015
3. Biro, M., Mashkoor, A., Sametinger, J., Seker, R. (eds.): Software safety and security risk mitigation in cyber-physical systems. IEEE Softw. **35**(1), 24–29 (2018)
4. Denning, T., Kohno, T., Levy, H.M.: Computer security and the modern home. Commun. ACM **56**(1), 94–103 (2013)
5. Doan, T.T., Safavi-Naini, R., Li, S., Avizheh, S., Venkateswarlu K., M., Fong, P.W.L.: Towards a resilient smart home. In: Proceedings of the 2018 Workshop on IoT Security and Privacy - IoT S&P 2018, Budapest, Hungary, pp. 15–21 (2018). https://dl.acm.org/citation.cfm?id=3229570
6. Eichelberger, F.: Using software defined radio to attack "smart home" systems. SANS Intitute (2015). https://www.sans.org/reading-room/whitepapers/threats/software-defined-radio-attack-smart-home-systems-35922
7. McGraw, G.: Software security. IEEE Secur. Priv. **2**(2), 80–83 (2004). https://doi.org/10.1109/MSECP.2004.1281254
8. Fernandes, E., Jung, J., Prakash, A.: Security analysis of emerging smart home applications. In: Proceedings of 37th IEEE Symposium on Security and Privacy, May 2016
9. Fisher, D.: Pair of bugs open honeywell home controllers up to easy hacks. https://threatpost.com/pair-of-bugs-open-honeywell-home-controllers-up-to-easy-hacks/113965/
10. c't press release: Weak Security In Loxone Smart Home System Lets Burglars Walk Right In, 31 August 2016. http://heise-gruppe.de/-3308694
11. Komninos, N., Lymberopoulos, D., Mantas, G.: Chapter 10 - Security in smart home environment. In: Wireless Technologies for Ambient Assisted Living and Healthcare: Systems and Applications, Medical Information Science Reference, USA, pp. 170–191 (2011)
12. The MITRE Corporation: Common Weakness Enumeration, A Community-Developed Dictionalry of Software Weakness Types. https://cwe.mitre.org
13. The MITRE Corporation: Common Vulnerabilities and Exposures, The Standard for Information Security Vulnerability Names. https://cve.mitre.org
14. Pretz, K.: IEEE provides the keys to a smarter home. Special Report on Smart Homes, 01 December 2015. http://theinstitute.ieee.org/technology-focus/technology-topic/ieee-provides-the-keys-to-a-smarter-home
15. ur Rehman, S., Gruhn, V.: An approach to secure smart homes in cyber-physical systems/Internet-of-Things. In: Fifth International Conference on Software Defined Systems (SDS), Barcelona, pp. 126–129 (2018)
16. Sametinger, J., Rozenblit, J.: Security scores for medical devices. In: SmartMedDev 2016 - Smart Medical Devices - From Lab to Clinical Practice, in Proceedings of the 9th International Joint Conference on Biomedical Engineering Systems and Technologies (BIOSTEC 2016) - Volume 5: HEALTHINF, Rome, Italy, 21–23 February 2016, pp. 533–541. ISBN 978-989-758-170-0
17. Sametinger, J., Steinwender, C.: Resilient medical device security. In: International Conference on Computational Science and Computational Intelligence, Symposium on Health Informatics and Medical Systems (CSCI-ISHI), NV, USA, Las Vegas, pp. 1775–1778 (2017)
18. Sasse, M.A., Smith, M. (eds.): The security-usability tradeoff myth. IEEE Secur. Priv. **14**(5), 11–13 (2016)

Deriving an Optimal Noise Adding Mechanism for Privacy-Preserving Machine Learning

Mohit Kumar[1,2](✉), Michael Rossbory[2], Bernhard A. Moser[2], and Bernhard Freudenthaler[2]

[1] Faculty of Computer Science and Electrical Engineering, University of Rostock, Rostock, Germany
`mohit.kumar@uni-rostock.de`
[2] Software Competence Center Hagenberg, Hagenberg, Austria
{`Michael.Rossbory,Bernhard.Moser,Bernhard.Freudenthaler`}`@scch.at`

Abstract. Differential privacy is a standard mathematical framework to quantify the degree to which individual privacy in a statistical dataset is preserved. We derive an optimal (ϵ, δ)–differentially private noise adding mechanism for real-valued data matrices meant for the training of models by machine learning algorithms. The aim is to protect a machine learning algorithm from an adversary who seeks to gain an information about the data from algorithm's output by perturbing the value in a sample of the training data. The fundamental issue of trade-off between privacy and utility is addressed by presenting a novel approach consisting of three steps: (1) the sufficient conditions on the probability density function of noise for (ϵ, δ)–differential privacy of a machine learning algorithm are derived; (2) the noise distribution that, for a given level of entropy, minimizes the expected noise magnitude is derived; (3) using entropy level as the design parameter, the optimal entropy level and the corresponding probability density function of the noise are derived.

Keywords: Privacy · Noise adding mechanism · Machine learning

1 Introduction

The data on which a machine learning or a data analytics algorithm operates might be owned by more than one party and a party may be unwilling to share its real data. The reason being that an algorithm's output may result in a leakage of private or sensitive information regarding the data. Differential privacy [3,5] is a standard framework to quantify the degree to which the data privacy of

The research reported in this paper has been partly supported by EU Horizon 2020 Grant 826278 "Serums" and the Austrian Ministry for Transport, Innovation and Technology, the Federal Ministry for Digital and Economic Affairs, and the Province of Upper Austria in the frame of the COMET center SCCH.

G. Anderst-Kotsis et al. (Eds.): DEXA 2019 Workshops, CCIS 1062, pp. 108–118, 2019.
https://doi.org/10.1007/978-3-030-27684-3_15

each individual in the dataset is preserved while releasing the algorithm output. Differential privacy is a property of an algorithm's data access mechanism and remains immune to any post-processing on the output of the algorithm. Machine learning methods such as deep neural networks have delivered remarkable results in a wide range of application domains. However, their training requires large datasets which might be containing sensitive information that need to be be protected from *model inversion* attack [6] and such issues have been addressed within the framework of differential privacy [1, 14].

The classical approach for attaining differential privacy for a real-valued function, where the function represents mathematically a machine learning algorithm, is to perturb the function output via adding noise calibrated to the global *sensitivity* of the function [4]. Adding of required amount (for attaining a given level of privacy) of noise would result in a loss of algorithm's accuracy and thus it is important to study the trade-off between privacy and accuracy [2, 10]. A general framework to provide utility guarantees for a single count query, subject to ϵ–differential privacy, was studied in [11]. A similar study taking a minimax model of utility for information consumers has been made in [12]. For single real-valued query function, a staircase-shaped probability density function was suggested in [8] for an optimal ϵ–differentially private noise adding mechanism. The approach was extended to the vector real-valued query function in [7]. For integer-valued query functions, the optimal mechanisms in (ϵ, δ)–differential privacy were studied in [9]. For single real-valued query function, the trade-off between privacy and utility in $(0, \delta)$–differential privacy was studied in [10].

Despite the fact that random noise adding mechanism has been widely used in privacy-preserving machine learning via output perturbation, there remains the challenge of studying privacy-utility trade-off for the algorithms performing a learning of the models with the matrix data (where e.g. rows corresponds to features and columns corresponds to samples). The aim is to protect a machine learning algorithm from an adversary who seeks to gain an information about the data from algorithm's output by perturbing the value in an element of the training data matrix. There is no standard approach to optimally design (ϵ, δ)–differentially private noise adding mechanism for real-valued data matrices used by a machine learning algorithm for the model training purpose. This study fills this gap by providing a general random noise adding mechanism for real-valued data matrices such that the mechanism, subject to (ϵ, δ)–differential privacy of a machine learning algorithm, minimizes the expected noise magnitude. To the best knowledge of authors, this is the first study of its kind to provide entropy based approach for resolving the privacy-utility trade-off for real-valued data matrices.

2   Sufficient Conditions for Differential Privacy

Consider a dataset consisting of N number of samples with each sample having p number of attributes. Assuming the data as numeric, the dataset can be represented by a matrix, say $Y \in \mathbb{R}^{p \times N}$. The machine learning algorithms typically train a model using available dataset. A given machine learning algorithm, training a model using data matrix Y, can be represented by a mapping, $\mathcal{A} : \mathbb{R}^{p \times N} \rightarrow \mathbf{M}$, where $\mathbf{M}$ is the model space. That is, for a given dataset Y, the algorithm builds a model $\mathcal{M} \in \mathbf{M}$ such that $\mathcal{M} = \mathcal{A}(Y)$. The privacy of data can be preserved via adding a suitable random noise to data matrix before the application of algorithm $\mathcal{A}$ on the dataset. This will result in a private version of algorithm $\mathcal{A}$ which is formally defined by Definition 1.

Definition 1 (A Private Algorithm on Data Matrix). *Let* $\mathcal{A}^+ : \mathbb{R}^{p \times N} \rightarrow Range(\mathcal{A}^+)$ *be a mapping defined as*

$$\mathcal{A}^+ (Y) = \mathcal{A} (Y + V), \ V \in \mathbb{R}^{p \times N} \tag{1}$$

where V *is a random noise matrix with* $f_{v_j^i}(v)$ *being the probability density function of its* (j, i)*–th element* v_j^i*;* v_j^i *and* $v_j^{i'}$ *are independent from each other for* $i \neq i'$*; and* $\mathcal{A} : \mathbb{R}^{p \times N} \rightarrow \mathbf{M}$ *(where* $\mathbf{M}$ *is the model space) is a given mapping representing a machine learning algorithm. The range of* $\mathcal{A}^+$ *is as*

$$Range(\mathcal{A}^+) = \left\{ \mathcal{A} (Y + V) \mid Y \in \mathbb{R}^{p \times N}, V \in \mathbb{R}^{p \times N} \right\}. \tag{2}$$

We intend to protect the algorithm $\mathcal{A}^+$ from an adversary who seeks to gain an information about the data from algorithm's output by perturbing the values in a sample of the dataset. We seek to attain differential privacy for algorithm $\mathcal{A}^+$ against the perturbation in an element of Y, say (j_0, i_0)–th element, such that magnitude of the perturbation is upper bounded by a scalar d. The d–adjacency [13] definition for two real matrices is provided in Definition 2.

Definition 2 (*d*–Adjacency for Data Matrices). *Two matrices* $Y, Y' \in \mathbb{R}^{p \times N}$ *are d–adjacent if for a given* $d \in \mathbb{R}_+$*, there exist* $i_0 \in \{1, 2, \cdots, N\}$ *and* $j_0 \in \{1, 2, \cdots, p\}$ *such that* $\forall i \in \{1, 2, \cdots, N\}, j \in \{1, 2, \cdots, p\}$,

$$|y_j^i - y_j'^i| \leq \begin{cases} d, \ if \ i = i_0, j = j_0 \\ 0, \quad otherwise \end{cases}$$

where y_j^i *and* $y_j'^i$ *denote the* (j, i)*–th element of* Y *and* Y' *respectively. Thus,* Y *and* Y' *differ by only one element and the magnitude of the difference is upper bounded by* d.

Definition 3 ((ϵ, δ)–Differential Privacy for $\mathcal{A}^+$). *The algorithm* $\mathcal{A}^+ (Y)$ *is* (ϵ, δ)–*differentially private if*

$$Pr\{\mathcal{A}^+ (Y) \subset \mathcal{O}\} \leq \exp(\epsilon) Pr\{\mathcal{A}^+ (Y') \subset \mathcal{O}\} + \delta \tag{3}$$

for any measurable set $\mathcal{O} \subseteq Range(\mathcal{A}^+)$ *and for d–adjacent matrices pair* (Y, Y').

Result 1 (Sufficient Conditions for (ϵ, δ)–Differential Privacy). *The following conditions on the probability density function of noise $v_j^i \in \mathbb{R}$ are sufficient to attain (ϵ, δ)–differential privacy by algorithm $\mathcal{A}^+$ (Definition 1):*

$$\int_{\Theta} f_{v_j^i}(v) \, dv \geq 1 - \delta, \quad where \tag{4}$$

$$\Theta \stackrel{\text{def}}{=} \left\{ v \mid \sup_{\hat{d} \in [-d,d]} \frac{f_{v_j^i - \hat{d}}(v)}{f_{v_j^i}(v)} \leq \exp(\epsilon), \ f_{v_j^i}(v) \neq 0, \ v_j^i \in \mathbb{R} \right\}. \tag{5}$$

Proof. The proof follows an approach similar to that of [13]. Define a set $\mathbf{S} \subseteq \mathbb{R}^{p \times N}$ as

$$\mathbf{S} = \{ Y + V \mid \mathcal{A}(Y + V) \in \mathcal{O} \}. \tag{6}$$

Further, define $\mathbf{S}_j^i \subseteq \mathbb{R}$ as the set of (j, i)–th elements of members in $\mathbf{S}$, i.e.,

$$\mathbf{S}_j^i = \{ y_j^i + v_j^i \mid \mathcal{A}(Y + V) \in \mathcal{O} \}. \tag{7}$$

We have

$$Pr\{\mathcal{A}^+(Y) \in \mathcal{O}\} = Pr\{\mathcal{A}(Y + V) \in \mathcal{O}\} \tag{8}$$

$$= Pr\{Y + V \in \mathbf{S}\} \tag{9}$$

$$= \prod_{j=1}^{p} \prod_{i=1}^{N} Pr\{y_j^i + v_j^i \in \mathbf{S}_j^i\}. \tag{10}$$

Considering Y and Y' as d–adjacent matrices, there must exist an index, say (j_0, i_0), at which Y and Y' differ in value. Equality (10) can be expressed as

$$Pr\{\mathcal{A}^+(Y) \in \mathcal{O}\} = Pr\{y_{j_0}^{i_0} + v_{j_0}^{i_0} \in \mathbf{S}_{j_0}^{i_0}\} \prod_{j,i,j \neq j_0, i \neq i_0} Pr\{y_j^i + v_j^i \in \mathbf{S}_j^i\} \tag{11}$$

Now, consider

$$Pr\{y_{j_0}^{i_0} + v_{j_0}^{i_0} \in \mathbf{S}_{j_0}^{i_0}\}$$
$$= Pr\{y_{j_0}^{i_0} + v_{j_0}^{i_0} \in \mathbf{S}_{j_0}^{i_0} \mid v_{j_0}^{i_0} \in (\mathbb{R} \setminus \Theta)\} + Pr\{y_{j_0}^{i_0} + v_{j_0}^{i_0} \in \mathbf{S}_{j_0}^{i_0} \mid v_{j_0}^{i_0} \in \Theta\}. \tag{12}$$

It follows from the d–adjacency that there exists a $\hat{d} \in [-d, d]$ such that

$$y_{j_0}^{i_0} = y_{j_0}'^{i_0} - \hat{d}.$$

Thus,

$$Pr\{y_{j_0}^{i_0} + v_{j_0}^{i_0} \in \mathbf{S}_{j_0}^{i_0}\}$$
$$= Pr\{y_{j_0}'^{i_0} - \hat{d} + v_{j_0}^{i_0} \in \mathbf{S}_j^{i_0} \mid v_{j_0}^{i_0} \in (\mathbb{R} \setminus \Theta)\} + Pr\{y_{j_0}'^{i_0} - \hat{d} + v_{j_0}^{i_0} \in \mathbf{S}_{j_0}^{i_0} \mid v_{j_0}^{i_0} \in \Theta\}$$
$$= \int_{\{y_{j_0}'^{i_0} - \hat{d} + v_{j_0}^{i_0} \mid v_{j_0}^{i_0} \in (\mathbb{R} \setminus \Theta)\} \cap \mathbf{S}_{j_0}^{i_0}} f_{y_{j_0}'^{i_0} - \hat{d} + v_{j_0}^{i_0}}(v) \, dv$$
$$+ \int_{\{y_{j_0}'^{i_0} - \hat{d} + v_{j_0}^{i_0} \mid v_{j_0}^{i_0} \in \Theta\} \cap \mathbf{S}_{j_0}^{i_0}} f_{y_{j_0}'^{i_0} - \hat{d} + v_{j_0}^{i_0}}(v) \, dv. \tag{13}$$

Now, we derive upper bounds on both terms at the right hand side of (13). First, consider

$$\int_{\{y_{j_0}^{'io}-\hat{d}+v_{j_0}^{io} \mid v_{j_0}^{io} \in (\mathbb{R} \setminus \Theta)\} \cap \mathbf{S}_{j_0}^{io}} f_{y_{j_0}^{'io}-\hat{d}+v_{j_0}^{io}}(v) \ dv$$

$$\leq \int_{\{y_{j_0}^{'io}-\hat{d}+v_{j_0}^{io} \mid v_{j_0}^{io} \in (\mathbb{R} \setminus \Theta)\}} f_{y_{j_0}^{'io}-\hat{d}+v_{j_0}^{io}}(v) \ dv \tag{14}$$

$$= \int_{\mathbb{R} \setminus \Theta} f_{v_{j_0}^{io}}(v) \ dv \tag{15}$$

$$= 1 - \int_{\Theta} f_{v_{j_0}^{io}}(v) \ dv. \tag{16}$$

It follows from the definition of Θ, i.e. from (5), that

$$\int_{\{y_{j_0}^{'io}-\hat{d}+v_{j_0}^{io} \mid v_{j_0}^{io} \in \Theta\} \cap \mathbf{S}_{j_0}^{io}} f_{y_{j_0}^{'io}-\hat{d}+v_{j_0}^{io}}(v) \ dv$$

$$\leq \exp(\epsilon) \int_{\{y_{j_0}^{'io}-\hat{d}+v_{j_0}^{io} \mid v_{j_0}^{io} \in \Theta\} \cap \mathbf{S}_{j_0}^{io}} f_{y_{j_0}^{'io}+v_{j_0}^{io}}(v) \ dv \tag{17}$$

$$\leq \exp(\epsilon) \int_{\mathbf{S}_{j_0}^{io}} f_{y_{j_0}^{'io}+v_{j_0}^{io}}(v) \ dv \tag{18}$$

$$= \exp(\epsilon) Pr\{y_{j_0}^{'io} + v_{j_0}^{io} \in \mathbf{S}_{j_0}^{io}\}. \tag{19}$$

Using (16) and (19) in (13), we have

$$Pr\{y_{j_0}^{io} + v_{j_0}^{io} \in \mathbf{S}_{j_0}^{io}\} \leq 1 - \int_{\Theta} f_{v_{j_0}^{io}}(v) \ dv + \exp(\epsilon) Pr\{y_{j_0}^{'io} + v_{j_0}^{io} \in \mathbf{S}_{j_0}^{io}\}. \tag{20}$$

Under condition (4), inequality (20) leads to

$$Pr\{y_{j_0}^{io} + v_{j_0}^{io} \in \mathbf{S}_{j_0}^{io}\} \leq \delta + \exp(\epsilon) Pr\{y_{j_0}^{'io} + v_{j_0}^{io} \in \mathbf{S}_{j_0}^{io}\}. \tag{21}$$

Using (21) in (11), we have

$$Pr\{\mathcal{A}^+(Y) \in \mathcal{O}\}$$
$$\leq \delta \prod_{j,i,j \neq j_0, i \neq io} Pr\{y_j^i + v_j^i \in \mathbf{S}_j^i\}$$
$$+ \exp(\epsilon) Pr\{y_{j_0}^{'io} + v_{j_0}^{io} \in \mathbf{S}_{j_0}^{io}\} \prod_{j,i,j \neq j_0, i \neq io} Pr\{y_j^i + v_j^i \in \mathbf{S}_j^i\} \tag{22}$$

$$\leq \delta + \exp(\epsilon)Pr\{y'^{i_0}_{j_0} + v^{i_0}_{j_0} \in S^{i_0}_{j_0}\} \prod_{j,i,j\neq j_0, i\neq i_0} Pr\{y^i_j + v^i_j \in S^i_j\} \qquad (23)$$

$$= \delta + \exp(\epsilon)Pr\{y'^{i_0}_{j_0} + v^{i_0}_{j_0} \in S^{i_0}_{j_0}\} \prod_{j,i,j\neq j_0, i\neq i_0} Pr\{y'^i_j + v^i_j \in S^i_j\} \qquad (24)$$

$$= \delta + \exp(\epsilon)Pr\{Y' + V \in S\} \qquad (25)$$

$$= \delta + \exp(\epsilon)Pr\{\mathcal{A}(Y' + V) \in \mathcal{O}\} \qquad (26)$$

$$= \delta + \exp(\epsilon)Pr\{\mathcal{A}^+(Y') \in \mathcal{O}\}. \qquad (27)$$

That is, the condition (3) is satisfied and hence the result is proved.

Remark 1 (Sufficient Conditions for ϵ–Differential Privacy). The sufficient conditions for ϵ–differential privacy follow from (4) with $\delta = 0$ as

$$\int_{\Theta} f_{v^i_j}(v) \, dv = 1, \qquad (28)$$

$$\Theta = \left\{ v \mid \sup_{\hat{d}\in[-d,d]} \frac{f_{v^i_j - \hat{d}}(v)}{f_{v^i_j}(v)} \leq \exp(\epsilon), \ f_{v^i_j}(v) \neq 0, \ v^i_j \in \mathbb{R} \right\}. \qquad (29)$$

where Θ is defined as in (5). The equality in (28) is due to the fact that the integral of any probability density function over a subset can't exceed unity.

3   An Optimal Differentially Private Noise

Result 2 (Minimum Magnitude for a Given Entropy Level). *The probability density function of noise that, for a given level of entropy, minimizes the expected noise magnitude is given as*

$$f^*_{v^i_j}(v; h) = \frac{1}{\exp(h-1)} \exp\left(-\frac{2|v|}{\exp(h-1)}\right), \qquad (30)$$

where h is the given entropy level. The expected noise magnitude is given as

$$E_{f^*_{v^i_j}}[|v|](h) = \frac{1}{2}\exp(h-1). \qquad (31)$$

Proof. We seek to solve

$$f_{v_j^i}^*(v; h) = \arg \min_{f_{v_j^i}(v)} \int_{\mathbb{R}} |v| f_{v_j^i}(v) \, dv \tag{32}$$

subject to

$$\int_{\mathbb{R}} f_{v_j^i}(v) \, dv = 1 \tag{33}$$

$$-\int_{\mathbb{R}} \log\left(f_{v_j^i}(v)\right) f_{v_j^i}(v) \, dv = h. \tag{34}$$

Introducing Lagrange multiplier λ_1 for (33) and λ_2 for (34), the following Lagrangian is obtained:

$$\mathcal{L}(f_{v_j^i}, \lambda_1, \lambda_2) = \int_{\mathbb{R}} |v| f_{v_j^i}(v) \, dv + \lambda_1 \left(\int_{\mathbb{R}} f_{v_j^i}(v) \, dv - 1 \right)$$
$$+ \lambda_2 \left(h + \int_{\mathbb{R}} \log\left(f_{v_j^i}(v)\right) f_{v_j^i}(v) \, dv \right).$$

The functional derivative of $\mathcal{L}$ with respect to $f_{v_j^i}$ is given as

$$\frac{\delta \mathcal{L}}{\delta f_{v_j^i}} = |v| + \lambda_1 + \lambda_2 \left(1 + \log\left(f_{v_j^i}(v)\right) \right). \tag{35}$$

Setting $\delta \mathcal{L}/\delta f_{v_j^i}$ equal to zero, we have

$$f_{v_j^i}(v) = \exp(-1 - \frac{\lambda_1}{\lambda_2}) \exp(-\frac{|v|}{\lambda_2}), \; \lambda_2 \neq 0. \tag{36}$$

Setting $\partial \mathcal{L}/\partial \lambda_1$ equal to zero and then solving using (36), we get

$$f_{v_j^i}(v) = \frac{1}{2\lambda_2} \exp(-\frac{|v|}{\lambda_2}), \; \lambda_2 > 0. \tag{37}$$

Setting $\partial \mathcal{L}/\partial \lambda_2$ equal to zero and then solving using (37), we get the optimal value of λ_2 as

$$\lambda_2^* = \frac{1}{2} \exp(h - 1). \tag{38}$$

Using the optimal value of λ_2^* in (37), the optimal expression for $f_{v_j^i}(v)$ is obtained as in (30). As $\lambda_2^* > 0$, $\mathcal{L}$ is convex in $f_{v_j^i}$ and thus $f_{v_j^i}^*$ corresponds to the minimum. Finally, the expected noise magnitude for $f_{v_j^i}^*$ is given by (31).

Result 3 (An Optimal ϵ–Differentially Private Noise). *The probability density function of noise that minimizes the expected noise magnitude together with satisfying the sufficient conditions for ϵ–differential privacy is given as*

$$f_{v_j^i}^*(v) = \frac{\epsilon}{2d} \exp(-\frac{\epsilon}{d}|v|). \tag{39}$$

The optimal value of expected noise magnitude is given as

$$E_{f_{v_j^i}^*}[\|v\|] = \frac{d}{\epsilon}. \tag{40}$$

Proof. Let h^* be the entropy of the optimal probability density function of the noise satisfying the sufficient condition for ϵ–differential privacy. It follows from Result 2 that the expression for optimal probability density function is given as

$$f_{v_j^i}^*(v; h^*) = \frac{1}{\exp(h^* - 1)} \exp(-\frac{2|v|}{\exp(h^* - 1)}). \tag{41}$$

Now, we have

$$\sup_{\hat{d} \in [-d,d]} \frac{f_{v_j^i - \hat{d}}^*(v; h^*)}{f_{v_j^i}^*(v; h^*)} = \exp(\frac{2d}{\exp(h^* - 1)}). \tag{42}$$

Since $f_{v_j^i}^*(v; h^*)$ satisfies the sufficient conditions (28–29), we have

$$\exp(\frac{2d}{\exp(h^* - 1)}) \leq \exp(\epsilon). \tag{43}$$

That is,

$$\frac{1}{2} \exp(h^* - 1) \geq \frac{d}{\epsilon}. \tag{44}$$

The left hand side of (44) is equal to the expected noise magnitude for $f_{v_j^i}^*(v; h^*)$. That is,

$$E_{f_{v_j^i}^*}[\|v\|](h^*) \geq \frac{d}{\epsilon}. \tag{45}$$

It follows from (45) that the minimum possible value of expected noise magnitude is equal to the right hand side of (45). The value of h^*, resulting in the minimum expected noise magnitude, is given as

$$h^* = 1 + \log\left(2\frac{d}{\epsilon}\right). \tag{46}$$

The value of h^* is put into (41) to obtain (39). The optimal density function (39) satisfies the sufficient conditions (28–29) for $\Theta = \mathbb{R}$.

Result 3 justifies the widely used Laplacian distribution for ϵ–differential privacy.

Result 4 (An Optimal (ϵ, δ)–Differentially Private Noise). *The probability density function of noise that minimizes the expected noise magnitude together with satisfying the sufficient conditions for (ϵ, δ)–differential privacy is given as*

$$f^*_{v^i_j}(v) = \begin{cases} \delta Dirac\delta(v), & v = 0 \\ (1-\delta)\frac{\epsilon}{2d}\exp(-\frac{\epsilon}{d}|v|), & v \in \mathbb{R} \setminus \{0\} \end{cases} \tag{47}$$

where $Dirac\delta(v)$ is Dirac delta function satisfying $\int_{-\infty}^{\infty} Dirac\delta(v)\ dv = 1$. The optimal value of expected noise magnitude is given as

$$E_{f^*_{v^i_j}}[|v|] = (1-\delta)\frac{d}{\epsilon}. \tag{48}$$

Proof. It is obvious that the optimal noise density function (39) satisfies the sufficient conditions (4–5) with $\Theta = \mathbb{R}$ for any $\delta \in [0,1]$ and thus attain (ϵ, δ)–differential privacy for any $\delta \in [0,1]$. However, in this case (i.e. when $\Theta = \mathbb{R}$ and $\delta > 0$), the lower bound on $\int_{\Theta} f_{v^i_j}(v)\ dv$ in (4) is not tight. Therefore, we need to derive an optimal density function for (ϵ, δ)–differential privacy taking $\Theta \subset \mathbb{R}$. Let $v_0 \in \mathbb{R}$ be a point which is excluded from $\mathbb{R}$ to define Θ, i.e.,

$$\Theta = \mathbb{R} \setminus \{v_0\}. \tag{49}$$

We extend the solution space for optimization by considering the discontinuous distributions having an arbitrary probability mass r at an arbitrary point v_0. Let $f_{v^i_j}\left(v; v_0, r, q_{v^i_j}(v)\right)$ be an arbitrary density function defined as

$$f_{v^i_j}\left(v; v_0, r, q_{v^i_j}(v)\right) = \begin{cases} rDirac\delta(v - v_0), & v = v_0 \\ (1-r)q_{v^i_j}(v), & v \in \Theta \end{cases} \tag{50}$$

Here, $q_{v^i_j}(v)$ is an arbitrary density function with a continuous cumulative distribution function and satisfying the sufficient conditions (28–29) for ϵ–differential privacy. As $q_{v^i_j}(v)$ is an arbitrary density function, the expected noise magnitude for $q_{v^i_j}(v)$ must be greater than or equal to the optimal value (40), i.e.,

$$\int_{\mathbb{R}} |v|q_{v^i_j}(v)\ dv \geq \frac{d}{\epsilon} \tag{51}$$

$$\int_{\Theta} |v|q_{v^i_j}(v)\ dv + \underbrace{\int_{\{v_0\}} |v|q_{v^i_j}(v)\ dv}_{=0} \geq \frac{d}{\epsilon}. \tag{52}$$

Here, the integral over a single point is equal to zero because of a continuous cumulative distribution function associated to $q_{v^i_j}(v)$. Thus,

$$\int_{\Theta} |v|q_{v^i_j}(v)\ dv \geq \frac{d}{\epsilon}, \tag{53}$$

where equality occurs if $q_{\mathbf{v}_j^i}(v)$ is equal to (39). Also

$$\int_{\Theta} q_{\mathbf{v}_j^i}(v) \, dv = \int_{\mathbb{R}} q_{\mathbf{v}_j^i}(v) \, dv - \int_{\{v_0\}} q_{\mathbf{v}_j^i}(v) \, dv \tag{54}$$

$$= 1. \tag{55}$$

Thus

$$\int_{\Theta} f_{\mathbf{v}_j^i}\left(v; v_0, r, q_{\mathbf{v}_j^i}(v)\right) \, dv = 1 - r. \tag{56}$$

For the density function (50) to satisfy condition (4), we must have

$$1 - r \geq 1 - \delta. \tag{57}$$

The expected noise magnitude for the density function (50) is given as

$$E_{f_{\mathbf{v}_j^i}}\left[|v|\right](v_0, r, q_{\mathbf{v}_j^i}(v)) = r \underbrace{|v_0|}_{\geq 0} + \underbrace{(1-r)}_{\geq 1-\delta} \underbrace{\int_{\Theta} |v| q_{\mathbf{v}_j^i}(v) \, dv}_{\geq d/\epsilon}. \tag{58}$$

It follows immediately that $E_{f_{\mathbf{v}_j^i}}\left[|v|\right]$ is minimized together with satisfying the sufficient conditions (4–5) with the following optimal choices for $(v_0, r, q_{\mathbf{v}_j^i}(v))$: $v_0^* = 0$, $r^* = \delta$, and $q_{\mathbf{v}_j^i}^*(v) = \frac{\epsilon}{2d} \exp(-\frac{\epsilon}{d}|v|)$. The result is proved after putting the optimal values into (50).

4   Concluding Remarks

This paper has stated an approach to derive an optimal (ϵ, δ)–differentially private noise adding mechanism for privacy-preserving machine learning. This is the first study to address the fundamental issue of trade-off between privacy and utility for matrix-valued query functions. Using noise entropy level as a design parameter for resolving the privacy-utility trade-off is a novel idea that would be further explored in our future work to link differential privacy with information-theoretic machine learning.

References

1. Abadi, M., et al.: Deep learning with differential privacy. In: Proceedings of the 2016 ACM SIGSAC Conference on Computer and Communications Security, CCS 2016, pp. 308–318. ACM, New York (2016)
2. Balle, B., Wang, Y.: Improving the Gaussian mechanism for differential privacy: analytical calibration and optimal denoising. CoRR abs/1805.06530 (2018)
3. Dwork, C., Kenthapadi, K., McSherry, F., Mironov, I., Naor, M.: Our data, ourselves: privacy via distributed noise generation. In: Vaudenay, S. (ed.) EUROCRYPT 2006. LNCS, vol. 4004, pp. 486–503. Springer, Heidelberg (2006). https://doi.org/10.1007/11761679_29

4. Dwork, C., McSherry, F., Nissim, K., Smith, A.: Calibrating noise to sensitivity in private data analysis. In: Halevi, S., Rabin, T. (eds.) TCC 2006. LNCS, vol. 3876, pp. 265–284. Springer, Heidelberg (2006). https://doi.org/10.1007/11681878_14
5. Dwork, C., Roth, A.: The algorithmic foundations of differential privacy. Found. Trends Theor. Comput. Sci. **9**(3–4), 211–407 (2014)
6. Fredrikson, M., Jha, S., Ristenpart, T.: Model inversion attacks that exploit confidence information and basic countermeasures. In: Proceedings of the 22nd ACM SIGSAC Conference on Computer and Communications Security, CCS 2015, pp. 1322–1333. ACM, New York (2015)
7. Geng, Q., Kairouz, P., Oh, S., Viswanath, P.: The staircase mechanism in differential privacy. IEEE J. Sel. Topics Signal Process. **9**(7), 1176–1184 (2015)
8. Geng, Q., Viswanath, P.: The optimal noise-adding mechanism in differential privacy. IEEE Trans. Inf. Theory **62**(2), 925–951 (2016)
9. Geng, Q., Viswanath, P.: Optimal noise adding mechanisms for approximate differential privacy. IEEE Trans. Inf. Theory **62**(2), 952–969 (2016)
10. Geng, Q., Ding, W., Guo, R., Kumar, S.: Optimal noise-adding mechanism in additive differential privacy. CoRR abs/1809.10224 (2018)
11. Ghosh, A., Roughgarden, T., Sundararajan, M.: Universally utility-maximizing privacy mechanisms. SIAM J. Comput. **41**(6), 1673–1693 (2012)
12. Gupte, M., Sundararajan, M.: Universally optimal privacy mechanisms for minimax agents. In: Proceedings of the Twenty-Ninth ACM SIGMOD-SIGACT-SIGART Symposium on Principles of Database Systems, PODS 2010, pp. 135–146. ACM, New York (2010)
13. He, J., Cai, L.: Differential private noise adding mechanism: basic conditions and its application. In: 2017 American Control Conference (ACC), pp. 1673–1678, May 2017
14. Phan, N., Wang, Y., Wu, X., Dou, D.: Differential privacy preservation for deep auto-encoders: an application of human behavior prediction. In: Proceedings of the Thirtieth AAAI Conference on Artificial Intelligence, AAAI 2016, pp. 1309–1316. AAAI Press (2016)

Linking Trust to Cyber-Physical Systems

Dagmar Auer[1,2(✉)], Markus Jäger[3], and Josef Küng[1,2]

[1] Institute for Application-oriented Knowledge Processing (FAW),
Johannes Kepler University Linz (JKU), Linz, Austria
{dauer,jkueng}@faw.jku.at
[2] LIT Secure and Correct Systems Lab, Linz Institute of Technology (LIT),
Johannes Kepler University Linz (JKU), Linz, Austria
{dagmar.auer,josef.kueng}@jku.at
[3] Pro2Future GmbH, Altenberger Strasse 69, 4040 Linz, Austria
markus.jaeger@pro2future.at

Abstract. Cyber-physical systems deeply connect artefacts, systems
and people. As we will use these systems on a daily basis, the question arises to what extent we can trust them, and even more, whether
and how we can develop computational models for trust in such systems. Trust has been explored in a wide range of domains, starting with
philosophy via psychology, sociology, and economy. Also in automation,
computing and networking definitions, models and computations have
emerged. In this paper we discuss our initial considerations on linking
concepts and models of trust and cyber-physical systems. After structuring and selecting basic approaches in both areas, we propose appropriate
links, which can be used as action areas for future development of trust
models and computations in cyber-physical systems.

Keywords: Cyber-physical systems · Trust · Trust models · Linking

1 Introduction

Today, industry, business and politics are already setting high expectations in
cyber-physical systems (CPSs). CPSs can be used in a large variety of application domains such as manufacturing, energy production and distribution, smart
home, transportation, logistics, farming or healthcare. These systems can significantly improve both, the scope and the quality of services. Still, many people
today, even being fascinated by these systems, strongly mistrust or even fear
them [11]. But trust is not only a relevant factor for users of these systems,
but also within itself, as CPSs are strongly characterized by cooperation of distributed, heterogeneous entities, systems, and humans.

The term *cyber-physical system (CPS)* has first been used in 2006 [1,3,6,
14]. As CPS spans an immense class of systems and involves several fields of
science and engineering [12], there is not the one definition of CPS, but rather a
number of them, strongly driven by the respective backgrounds and application
domains [3].

© Springer Nature Switzerland AG 2019
G. Anderst-Kotsis et al. (Eds.): DEXA 2019 Workshops, CCIS 1062, pp. 119–128, 2019.
https://doi.org/10.1007/978-3-030-27684-3_16

In [6] a high-level definition is provided, which gives a first idea of CPS: "Cyber-physical systems (CPS) are smart systems that include engineered interacting networks of physical and computational components".

The discussion of CPS definitions in [12] is summarized as "... CPS are systems featuring a tight combination of, and coordination between network systems and physical systems. By organic integration and in-depth collaboration of computation, communications and control (3C) technology, they can realize the real-time sensing, dynamic control and information services of large engineering systems".

Both definitions are not taking humans explicitly into account. However, when studying the models, approaches, etc. [1,6,12], humans turn out to be an important part of the overall CPS (e.g., as system user [12], human and human aspect [1,6]). Several other definitions [3,17] explicitly emphasize humans and their roles within the deep collaboration in the CPS. We join in this view and consider the three components - human, cyber, and physical - in the following as equivalent, similar building blocks of the entire system.

The importance of trust in the context of CPSs and the many still open research questions have already been addressed in several publications [4,6,11, 15].

The rest of the paper is structured as follows: In Sect. 2 core characteristics of a CPS are revealed by discussing basic aspects of the "3C CPS Architecture" by Ahmadi et al. [1] and the "CPS Conceptual Model" by Griffor et al. [6]. The concept of trust and an overview of different trust models are given in Sect. 3. In Sect. 4, a much broader, but for now rather initial discussion of linking the concept of trust to core components of a CPS is presented. The paper concludes with a summary in Sect. 5, and gives an outlook on future issues and challenges.

2   Models of Cyber-Physical Systems

Based on the definitions of CPS in Sect. 1, the two models of CPS - "3C CPS Architecture" [1] and the "CPS Conceptual Model" [6] - will be discussed in more detail.

The *CPS Architecture for Industry 4.0* [1] stresses humans as one of its three core components - the *human component (HC)*, the *cyber component (CC)*, and the *physical component (PC)*. These components are connected by specific interfaces. Figure 1 gives a simplified version of the 3C CPS Architecture, with focus on the relevant aspects for the further discussion on linking trust to CPSs in Sect. 4. The core components and interfaces are characterized in [1] as follows.

Human Component (HC): Humans are an essential part of a CPS by providing data, needs, feedback, rules, etc. and by using the outcome. As the complexity of a CPS is typically much higher than with former systems, the demands on humans interacting with these systems are heavily increasing.

Cyber Component (CC): Software services and the overall cyber (component) architecture are subsumed by the term cyber component [1,14]. Services deal with data, information, and knowledge management and processing.

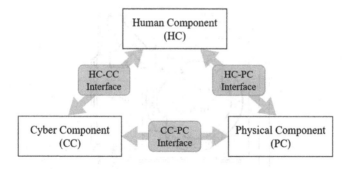

Fig. 1. 3C CPS Architecture (based on [1])

Physical Component (PC): A PC is a specific technological artefact, e.g., sensor or actuator, to interact with the lower level hardware parts. It provides measurement data or receives control data.

These three core components are integrated via interfaces. The HC-CC Interface bridges human and cyber components. Thus, Human Computer Interaction (HCI) aspects are particularly relevant in this context. The CC-PC Interface, which links cyber and physical components is much about data acquisition, data fusion, data integration, and hardware control. The physical components send data (e.g., monitoring, tracing, tracking data) to the cyber components to be collected, integrated, and processed, while these components provide adequate control data (e.g., automation or reconfiguration data) to the physical components. With the HC-PC Interface approaches from Human Machine Interface (HMI) or latest research areas like Robopsychology are becoming increasingly important.

The CPS Conceptual Model [6] also contains the three components - cyber, physical and human. Even though all relevant parts of the 3C CPS Architecture [1] can be found in Fig. 2, the focus differs. With [1] the human component takes an equivalent role in the overall system, such as the cyber and physical components, whereas [6] place humans in the CPS environment. [6] argue that "... humans function in a different way than the other components of a CPS". Humans need different roles, e.g., input provider, output user, controller of a CPS, or even object of a CPS, when they interact with a CPS [6].

The different view on humans in the context of a CPS is also clearly stated by the CPS definition [6]: "Cyber-physical systems integrate computation, communication, sensing, and actuation with physical systems to fulfill time-sensitive functions with varying degrees of interaction with the environment, including human interaction".

The model focuses on the interaction between the cyber and physical worlds. The physical state of the physical world is transmitted to the cyber world via information. This information is used in the cyber world to take decisions concerning action(s) in the physical world. This collaboration pattern is not limited to one level of the system, but it can be applied recursively. Thus, it can be used for a single device as well as for systems of systems.

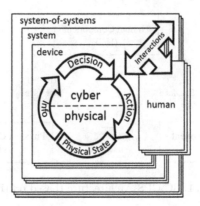

Fig. 2. CPS Conceptual Model [6]

We refer to these two models in our further discussion, as they clearly discuss the core components - physical, cyber and human - and their interactions [1,6]. Furthermore, [6] describe the recursive use of the basic model on different layers, from single devices to systems of systems.

3 Definitions and Models for Trust

The question of trusting people or subjects has been discussed for decades, gaining even more importance today. Besides common definitions of the term trust, selected models of trust are discussed in this section.

3.1 Definitions of Trust

People typically trust man-made technology - from cars to planes, computers to space shuttles. As long as they work properly, nobody even thinks about (not) trusting them. Only in case of problems, the question of trust arises.

Trust is a multidisciplinary concept with partly different meanings in different domains. Subjective probability, expectation, willingness, belief, or attitude are used to characterize trust. The following Table 1 gives a brief overview.

In the organizational context *trust between agents (or subjects)* is an important concept. It means the conviction of one person concerning the correctness of the statements of some other person or group. Trust depends on the corresponding persons or groups and can be influenced by past events and experiences [13]. Trust between agents is subjective.

In the context of IT, [2] define three main types of trust (based on [16]):

Trusting beliefs "means a secure conviction that the other party has favourable attributes (such as benevolence, integrity, and competence)" [2].

Trusting intentions "means a secure, committed willingness to depend upon, or to become vulnerable to, the other party in specific ways" [2].

Table 1. Multidisciplinary definitions of trust [5]

Discipline	Meaning of Trust
Sociology	Subjective probability that another party will perform an action that will not hurt my interest under uncertainty and ignorance
Philosophy	Risky action deriving from personal, moral relationships between two entities
Economics	Expectation upon a risky action under uncertainty and ignorance based on the calculated incentives for the action
Psychology	Cognitive learning process obtained from social experiences based on the consequences of trusting behaviors
Organizational Management	Willingness to take risk and being vulnerable to the relationship based on ability, integrity, and benevolence
International Relations	Belief that the other party is trustworthy with the willingness to reciprocate cooperation
Automation	Attitude that one agent will achieve another agent's goal in a situation where imperfect knowledge is given with uncertainty and vulnerability
Computing & Networking	Estimated subjective probability that an entity exhibits reliable behavior for particular operation(s) under a situation with potential risks

Trusting behavior is about manifesting the "willingness to depend" [2] on some other party.

The notion of trust heavily depends on the application domain and strongly influences the design of the models to determine trust.

3.2 Trust Models

Selected established trust models as well as the Weighted Arithmetic Mean Trust Model for trust propagation are briefly presented below.

One of the easiest ways to represent trust values is the *Binary Trust Model*. Trust values are represented either by 0 (fully trusting) or 1 (not trusting). The main challenge is to define the threshold between trusting and not trusting.

The *Probabilistic Trust Model* allows for more detailed trust values. Trust values are defined in the range of 0 to 1, often also as percentage values. Besides not trusting (trust = 0) and fully trusting (trust = 1) an entity, a (theoretically) infinite number of trust values can be defined. However, the problem of interpreting the trust value and defining thresholds still exists.

In the following more sophisticated models are introduced, which take several factors into account to determine trust.

Trust Model by Mayer. In this integrative model of organizational trust [13], trust is composed of ability, benevolence and integrity. Besides the tendency of the trustor to trust these factors and the overall trust level, also risk is considered (cp. Fig. 3).

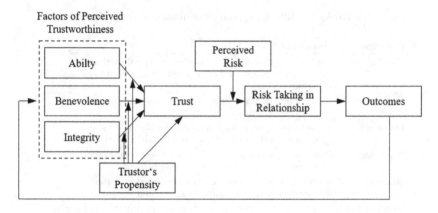

Factors of Perceived
Trustworthiness

Fig. 3. Integrated model of interpersonal trust [13]

This model is used in many domains, but has its roots in the field of management.

Opinion-Space Trust Model. This model by Jøsang [9,10] is well established. It contains an *evidence space* and an *opinion space*, which are two equivalent models for representing human beliefs, which can be summarized as trust. The opinion-space trust model (cp. Fig. 4) consists of *belief b*, *disbelief d*, and *uncertainty u* to represent trust. The sum of these three variables is always 1.

Weighted Arithmetic Mean (WAM) Trust Model. This model considers the propagation of trust and precision values. It introduces the factor importance with the calculation of the trust value. The initial trust value is assumed to be similar to the one in the probabilistic model, but the further processing differs. With the WAM trust model an importance-weighted arithmetic mean is calculated. The resulting trust value is always between 0 and 1. This trust model has been designed for multi-step knowledge processing systems, but also been tested in other scenarios [7,8].

In the next section the presented trust models are discussed in the context of CPSs.

4   Linking Trust to Cyber-Physical Systems

The following concept describes the current state of our ongoing work on linking trust to CPSs.

A CPS integrates different devices, systems or even systems of systems, typically as black boxes as they are not part of the same trust area. Furthermore, trust management is usually decentralized, therefore no overall agreed trust value is available. Thus, to determine the trust value, we need to rely on sources such

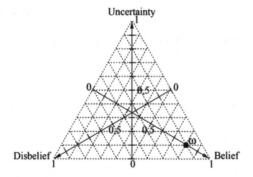

Fig. 4. Graphical description of the opinion-space trust model [9,10]

as the information given by the provider of these systems, information captured during earlier use, recommendations, reputation, and/or some additional evidence to compute an individual trust value.

Trust is always some directed relationship between two parties, e.g., A trusts B. Interaction in a CPS can be horizontal or vertical. The cyber, physical, and human components of the Three Components (3C) CPS Architecture [1] are integrated horizontally via specific interfaces, whereas the CPS Conceptual Model [6] focuses on recursive, thus, vertical integration of the components involved.

With the 3C CPS Architecture, interfaces are part of the overall model. These interfaces need to be considered from both directions, as trust is a unidirectional concept. Some of the trust models described in Sect. 3.2 such as the Binary Trust Model and Probabilistic Trust Model are generally applicable to all relations. Others [7,9,10,13] rather focus on humans, who apply concepts such as belief, disbelief, benevolence, integrity and/or interest to establish trust, but also consider computational models for determining trust between cyber and physical components. Specific trust models focusing on trust between cyber and physical components have not been considered so far in our work. The following Table 2 gives an overview of linking the interfaces discussed in Sect. 2 to selected trust models from Sect. 3.2.

For now, only trust between two components has been regarded. The overall level of trust always needs to be considered within the context of all relevant interacting partners and the concrete layout of the CPS. Thus, some kind of trust propagation is needed. The binary and proportional trust models allow for easy propagation of trust, while [7,10] show more sophisticated models.

To derive the overall trust level, horizontal as well as vertical integration are needed. Horizontal integration of trust means to take all trust values of all relevant interfaces with the interacting partners into account (cp. 3C CPS Architecture [1] in Sect. 3.2). Vertical integration of trust in contrast is about the influence of an integrated component's trust in its own interacting partners (cp. CPS Conceptual Model by [6] in Sect. 3.2). This strategy can be adapted recursively. A prerequisite is that each component is ready to propagate its

Table 2. Linking CPS component interfaces to trust models

Component interface	Binary/probab. trust model	Trust model by Mayer	Opinion-space trust model	WAM trust model
HC-CC human to cyber	X	X	X	X
CC-HC cyber to human	X			
HC-PC human to physical	X	X	X	X
PC-HC physical to human	X			
CC-PC cyber to physical	X		X	X
PC-CC physical to cyber	X		X	X

trust level to the referencing component. By this, the component-specific trust levels can be used besides other information (e.g., failure statistics, ratings) to calculate the trust level on the next higher layer of the overall system. This approach allows to propagate trust in CPSs with decentralized components in distinct trust areas.

This process of deriving statements about trusting the entire system needs to be evaluated according to trust propagation and aggregation methods. Here we see a variety of open research questions and options for better understanding trust in CPSs.

5 Conclusion

The domains of CPSs and trust have developed independently of each other, with a variety of definitions and models on both sides. Within this paper we show first considerations of how links between CPS and trust could be established. To prepare for our discussion, we presented selected recognized definitions and models in each of the domains. The 3C CPS Architecture Model [1] explicitly states the three core components - cyber component, physical component, and human component - plus their interfaces, while the CPS Conceptual Model [6] stresses the recursive character of the CPS model. With trust, we considered five different models to describe trust between two distinct entities. Four of them allow trust propagation as well. When linking these trust models to the area of CPS, not all of them appear to be suitable for all kinds of relevant interactions. Thinking towards application of trust models in a CPS, we have to consider different instances of different trust models with different trust values for each of the numerous interactions. Combining these trust estimations and calculations, i.e., propagating and aggregating trust assigned to components to a trust statement about the complete system, is a challenging open research aspect for CPSs. Our future work will deal in particular with more sophisticated trust propagation and aggregation models in CPSs.

Acknowledgement. This work has partially been supported by the FFG, Contract No. 854184: "Pro^{2}Future is funded within the Austrian COMET Program—Competence Centers for Excellent Technologies—under the auspices of the Austrian Federal Ministry of Transport, Innovation and Technology, the Austrian Federal Ministry for Digital and Economic Affairs and of the Provinces of Upper Austria and Styria. COMET is managed by the Austrian Research Promotion Agency FFG.", and by the LIT Secure and Correct Systems Lab funded by the State of Upper Austria.

References

1. Ahmadi, A., Sodhro, A.H., Cherifi, C., Cheutet, V., Ouzrout, Y.: Evolution of 3C cyber-physical systems architecture for industry 4.0. In: Borangiu, T., Trentesaux, D., Thomas, A., Cavalieri, S. (eds.) SOHOMA 2018. Studies in Computational Intelligence, vol. 803, pp. 448–459. Springer, Cham (2019). https://doi.org/10.1007/978-3-030-03003-2_35. https://www.springer.com/gp/book/9783030030025
2. Andersen, B., Kaur, B., Tange, H.: Computational trust. In: Khajuria, S., Bøgh Sørensen, L., Skouby, K.E. (eds.) Cybersecurity and Privacy - Bridging the Gap. River Publishers Series in Communications, pp. 83–98. River Publishers, Aalborg (2017)
3. Cardin, O.: Classification of cyber-physical production systems applications: proposition of an analysis framework. Comput. Ind. **104**, 11–21 (2019). https://doi.org/10.1016/j.compind.2018.10.002
4. Chang, E., Dillon, T.: Trust, reputation, and risk in cyber physical systems. In: Papadopoulos, H., Andreou, A.S., Iliadis, L., Maglogiannis, I. (eds.) AIAI 2013. IAICT, vol. 412, pp. 1–9. Springer, Heidelberg (2013). https://doi.org/10.1007/978-3-642-41142-7_1. https://www.springer.com/de/book/9783642411410
5. Cho, J.H., Chan, K., Adali, S.: A survey on trust modeling. ACM Comput. Surv. **48**(2), 1–40 (2015). https://doi.org/10.1145/2815595
6. Griffor, E.R., Greer, C., Wollman, D.A., Burns, M.J.: Framework for cyber-physical systems: volume 1, overview. National Institute of Standards and Technology, Gaithersburg, MD (2017). https://doi.org/10.6028/NIST.SP.1500-201
7. Jäger, M., Küng, J.: Introducing the factor importance to trust of sources and certainty of data in knowledge processing systems - a new approach for incorporation and processing. In: Proceedings of the 50th Hawaii International Conference on System Sciences, vol. 50, pp. 4298–4307. IEEE, January 2017. https://doi.org/10.24251/HICSS.2017.521. Accessed 07 Sept 2018
8. Jäger, M., Nadschläger, S., Küng, J.: Concepts for trust propagation in knowledge processing systems - a brief introduction and overview. In: Proceedings of the 17th IEEE International Conference on Trust, Security and Privacy in Computing and Communications (IEEE TrustCom-2018), pp. 1502–1505 (2018). https://doi.org/10.1109/TrustCom/BigDataSE.2018.00212
9. Jøsang, A., Knapskog, S.J.: A metric for trusted systems. In: Proceedings of the 21st NIST-NCSC National Information Systems Security Conference, pp. 16–29. NSA, August 1998. http://folk.uio.no/josang/papers/JK1998-NSC.pdf

10. Jøsang, A., Marsh, S., Pope, S.: Exploring different types of trust propagation. In: Stølen, K., Winsborough, W.H., Martinelli, F., Massacci, F. (eds.) iTrust 2006. LNCS, vol. 3986, pp. 179–192. Springer, Heidelberg (2006). https://doi.org/10.1007/11755593_14
11. Lee, I.: Trust management for cyber-physical systems, 19 June 2016
12. Liu, Y., Peng, Y., Wang, B., Yao, S., Liu, Z.: Review on cyber-physical systems. IEEE/CAA J. Automatica Sinica **4**(1), 27–40 (2017). https://doi.org/10.1109/JAS.2017.7510349
13. Mayer, R.C., Davis, J.H., Schoorman, F.D.: An integrative model of organizational trust. Acad. Manag. Rev. **20**(3), 709–734 (1995). https://doi.org/10.2307/258792
14. Monostori, L., et al.: Cyber-physical systems in manufacturing. CIRP Ann. **65**(2), 621–641 (2016). https://doi.org/10.1016/j.cirp.2016.06.005
15. Rein, A., Rieke, R., Jäger, M., Kuntze, N., Coppolino, L.: Trust establishment in cooperating cyber-physical systems. In: Bécue, A., Cuppens-Boulahia, N., Cuppens, F., Katsikas, S., Lambrinoudakis, C. (eds.) CyberICS/WOS-CPS -2015. LNCS, vol. 9588, pp. 31–47. Springer, Cham (2016). https://doi.org/10.1007/978-3-319-40385-4_3
16. Rousseau, D., Sitkin, S., Burt, R., Camerer, C.: Not so different after all: a cross-discipline view of trust. Acad. Manag. Rev. **23**(3), 393–404 (1998). https://doi.org/10.5465/AMR.1998.926617
17. Schuh, G., Potente, T., Varandani, R., Hausberg, C., Fränken, B.: Collaboration moves productivity to the next level. Procedia CIRP **17**, 3–8 (2014). https://doi.org/10.1016/j.procir.2014.02.037

Security Risk Mitigation of Cyber Physical Systems: A Case Study of a Flight Simulator

Maryam Zahid[1(✉)], Irum Inayat[2], Atif Mashkoor[3,4],
and Zahid Mehmood[5]

[1] Capital University of Science and Technology (C.U.S.T.), Islamabad, Pakistan
maryam.zahid@cust.edu.pk
[2] Software Engineering and Automation Lab (S.E.A.L.),
National University of Computer and Emerging Sciences, Islamabad, Pakistan
irum.inayat@nu.edu.pk
[3] Software Competence Center Hagenberg GmbH, Hagenberg, Austria
atif.mashkoor@scch.at
[4] Johannes Kepler University Linz, Linz, Austria
atif.mashkoor@jku.at
[5] University of Lahore, Islamabad, Pakistan
zahid.mehmood@ee.uol.edu.pk

Abstract. Avionics has seen a greatest shift in technology over the last two decades. The severity of the consequences resulting from a lack of risk management in avionics can be seen from recent incidents of unmanned aerial vehicles being hacked or in the hacking of vendor-controlled systems installed in commercial aircrafts. Over a million incidents related to security breaches at cyber layer have been recorded over the last decade, among which 350,000 cyber-attacks alone have taken place in the year 2018. Unfortunately, only a limited set of studies have been conducted on security risk management, particularly specific to avionics. In this article, we aim to identify, analyze and mitigate the security risks of 6 Degree of Freedom Flight Simulator. As a result, we identify 8 risks of level 3–4 as per the IEC 61508 standard. Further analysis of the identified risks yields in another 34 risks. We then mitigate the severity of the identified risks from level 4 to level 2 as per the IEC 61508 standard. The cryptosystem used for risk mitigation performed relatively faster as compared to some of the most recently proposed encryption schemes.

Keywords: Cyber Physical Systems (CPS) · Cyber-security ·
Risk identification · Risk assessment · Risk mitigation · Risk management ·
Cryptosystem

1 Introduction

Cyber Physical Systems (CPS) are the embedded systems having controlled, coordinated, constantly monitored, operations integrated by a core based on Internet of Things (IoT). The requirement of having an automated system with constant connectivity

This work has been partially supported by the Austrian Ministry for Transport, Innovation and Technology, the Federal Ministry of Science, Research and Economy, and the State of Upper Austria in the frame of the COMET center SCCH, and the LIT Secure and Correct Systems Lab funded by the State of Upper Austria.

© Springer Nature Switzerland AG 2019
G. Anderst-Kotsis et al. (Eds.): DEXA 2019 Workshops, CCIS 1062, pp. 129–138, 2019.
https://doi.org/10.1007/978-3-030-27684-3_17

among physical devices and its users has led to the operations of systems including communication systems, home appliances, automotive electronics, games, weapons, and Supervisory Control and Data Acquisition (SCADA) systems being mapped onto CPS. Among all such requirements, adequate provision of security is the main feature a CPS relies on to gain the trust of its users. Although CPS have been proven to be beneficial to users involved in critical operations, they have also brought forward major security and privacy concerns, eventually leading to a state where users lose trust in such systems [1].

Constant connectivity among various components of CPS, limited computation power, bandwidth, storage, and power consumption contribute to being the reasons behind the complex nature of CPS, and therefore cannot be integrated with traditional security protocols and mechanisms [1]. Implementing complex security requirements for CPS alone is not enough due to multiple factors making the system vulnerable to hackers. To make the system more secure and capable of adequately handling user privacy and security threats, there needs to be a proper implementation of risk analysis and mitigation techniques along with elicitation and modeling of the system's security requirements itself [1].

This paper discusses the application of security risk mitigation strategy, mitigating cyber-security risks targeting authentication, data-freshness, data-integrity, confidentiality, and non-repudiation at the application layer of CPS while reducing the impact of the security risks within CPS. In particular, this article discusses the application of the strategy on software systems related to the domain of avionics. For that, first the risks are identified using misuse cases, quantitative risk assessment is performed using fault tree analysis and Security Integrity Levels (SIL) defined in IEC 61508 standard [2] and recommended by the DO 178-C standard [3], and risk mitigation is performed using a hybrid cryptosystem based upon a combination of 3DES, RSA and CMCIAA encryption schemes.

Following the introduction of this paper, Sect. 2 provides an overview of the related work on the security in CPS and risk management processes adopted to minimize the impact of the security risks on the system. Since this article focuses on the application of an existing risk mitigation strategy on avionics, Sect. 3 presents the experimental design adopted. Section 4 presents results of the conducted experiment followed by the discussion on the obtained results in Sect. 5. Before concluding the paper, we also discuss the limitations of the conducted study in Sect. 6.

2 Literature Review

2.1 Security in Cyber-Physical Systems

CPS and SCADA-based systems have high priority security requirements with a vulnerability level specific to the nature of the operations intended. As far as, security requirements are concerned, confidentiality, integrity and availability are considered to be the three main security objectives in applications related to avionics. Thus the cyber–attacks on the related systems can also be classified according to the security objective targeted [4].

Security requirements in CPS are implemented based on the role of each layer of its architecture [5]. However, poor implementation of design protocols, cyber laws and regulations, risk management principles as well as poor maintenance of system also contribute to it [6]. The security threats to a CPS recorded over the years range from physical damages to loopholes in software system exposing it to the hackers [7]. Physical destruction, equipment theft, inside job and cyber-attacks such as privacy leakage, Man-in-the-Middle attack (MitM), spoofing, unauthorized access, data tempering, and system hijacking [8] are some of the most reported attacks on CPS in literature.

2.2 Risk Management for Cyber-Physical Systems

Risk Identification

Risk identification being the first step of risk management, involves identifying the ways a misuser can pose a threat to the software system or an undesired event, such as a natural disaster or an accident resulting in a complete or a sub-system failure by emitting abnormal behavior [9].

During risk identification, source, target, motive, attack vector, and potential consequences are the five common factors required to be identified against every threat to the CPS security [10]. According to a study conducted, risk identification techniques can be categorized as: anomaly-based, specification–based, signature–based, reputation–based instruction detection technique [11], signal–based, packet–based, proactive and hybrid schemes [12]. The ability of the misuse cases to depict the possible paths a hacker can adopt to manipulate the data processed for further actions by the system, allows us to not only identify the security requirements but also represents the possible risks in the software system to be mitigated [6].

Risk Assessment

This step analyzes the identified risks for their probability of occurrence and severity of impact it can have on the system's performance. Unlike in traditional IT security systems, CPS requires identification, assignment and calculation of assets, threats and vulnerabilities [7] while fully taking into consideration its characteristics to finding a suitable risk assessment method [13]. Simulations and models representing the attack designs can help better assess the security risks of CPS [11] providing not only the theoretical guidelines detecting attacks but can also help determine resilience controls. Risk strategies are effective when the risks are timely identified. Based on the level of detail required, resources available and the type of system under development, risk assessment approaches can be divided into two main categories namely qualitative approaches such as the risk matrix [14] and the quantitative approaches such as Fault Tree Analysis, Event Trees or Risk Graphs [21]. Most of the standards designed for the development of SCADA systems in particular vehicular CPS recommend the use of quantitative risk assessment techniques, in-specific Fault Tree Analysis, Event Trees or Risk Graphs, due to their ability to explicitly visualizing the relationship between the events and the causes leading to the system failure [21].

Risk Mitigation

The mitigation techniques can be classified on the basis of the architectural layer of CPS applied at, i.e., network layer mitigation, application layer or physical layer mitigation [12]. One way to reduce system vulnerability is to convert security requirements (such as data-integrity, data-confidentiality, authentication, confidentiality, non-repudiation and data-freshness) and possibility of their associated threats into design decisions throughout the software development life cycle [15]. Application of encryption schemes (such as AES-ECB combined with ECC (Elliptic Curve Cryptography) [16], AES-GCM embedded onto GPU [17], AES-CCM [18], and AES-CBC [19]), defined authentication principles, limited access controls, digital forensics [20], remote attestation of embedded devices, management of security incident and events, and intrusion detection can be used to develop a secure supporting infrastructure necessary to accurately store and transmit information to the appropriate application [20].

Since CPS mainly includes security critical and safety critical infrastructures [21]; implementation of data-freshness and non-repudiation is of crucial importance. The techniques proposed over the years for risk mitigation are least focused on mitigating risks related to data-freshness, non-repudiation, and data-integrity at an application layer of CPS. In most cases, the risk mitigation techniques proposed are applicable either at the physical layer or the network layer of CPS. Most of the proposed risk mitigation techniques also lack validation [22] especially in the domain of avionics.

3   Experimental Design

3.1   Case Study

A software system simulating 6 Degree of Freedom (6 DoF) motion of an aircraft was used to test the applicability of a risk mitigation strategy in the domain of avionics. A 6 DoF simulator is a software system designed to train commercial pilots on take–off, landing, and flying the plane before sending them off to fly the actual aircraft (see Fig. 1). The software is developed in LabView[1] by a research and development organization simulating the motion of an aircraft during an actual flight.

The software captures and presents the actions taken by the pilot during the entire simulation and stores them in the log file which can be used in future simulations run by that particular pilot. The simulator records the highest and lowest speeds; motion coordinates data, and geographical coordinates reached by aircraft during the simulation. During the simulation, the system constantly monitors the obstacle detection sensors and platform checking levels to maintain the safety measures to run any particular flight simulations.

---

[1] www.ni.com/labview.

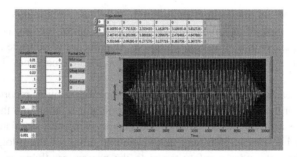

Fig. 1. 6 Degrees of Freedom Simulator

3.2 Attacks Simulated

Since we are focusing only on the security requirements and risks associated with them at an application layer, i.e., authentication, non–repudiation, data–integrity, data–freshness, and confidentiality we selected spoofing, data tempering and unauthorized access as the basic events to mitigate in this case study. For this, the case study was subjected to selective active and passive cyber-attacks namely: Man-in-the-Middle attack (MiTM) and spoofing. During these attacks, the attacker intercepts the data being transmitted between the various nodes of the simulator connected onto a network. The intercepted data consists of simulator's trajectories at a given time which if tempered with could mislead the system into violating safety conditions embedded into the code to stop specific set of motions not fit for indoor flight simulations (such as the barrel motion – requiring the hexapod to rotate to an angle $> \pm 70°$), thus causing causalities, heavy damage or major injuries. The reason for implementing these 2 attacks is that it allowed us to observe the severity of the system or subsystem failure in-terms of system being compromised, unauthorized access, a sequence of undesired events occurring, delayed operations, data-tempering and others.

3.3 Variables

In order to analyze the results of the experiment conducted, risk values and both the encryption and decryption time of the cryptosystem from our risk mitigation phase, were used as independent variables while the cryptosystem itself was used as a dependent variable.

4 Results

4.1 Security Requirement Extraction and Risk Identification

According to the misuse cases (see Figs. 2 and 3) the system starts the simulation based on the pilot's command. The actions performed by the pilot are continuously recorded by the system in the form of flight motion, while simulating the motions on the physical simulator. During the process the system checks for the platform levels, angular speeds reached during the simulation and the degrees to which the simulator is directed to position itself. In case of an invalid move such as the barrel motion, the system is to engage an emergency simulation halt operation. Also in case of obstacle detection in its surrounding, the system is to bring the simulator to an emergency halt for safety purposes.

The sensor readings related to the obstacle detection, motion recording, platform checking levels or the entire control of the flight simulator if fallen into the hands of the hacker can pose a threat to the lives of the people using the simulator. Consistent intrusion of the hacker can also lead to fallacious training of the pilot, which if not corrected on time may pose a safety threat to the passengers flying in a commercial under the captaincy of that particular pilot.

Therefore, the security requirements extracted from the misuse cases generated in general consist of; ensuring data and node authentication, and confidentiality of the data transmitted between the sensors and the actuators in the simulator, checking the freshness of the data received from the sensors and the control center before further processing it. The system should also check for non-repudiation in order to ensure a secure communicational channel between the simulator and the control center for sending appropriate commands under emergency situations.

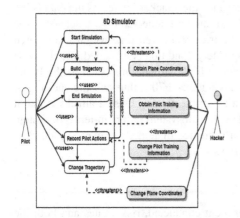

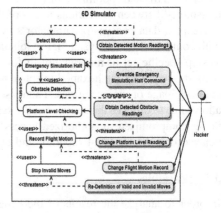

Fig. 2. Misuse case #1: flight motion recording

Fig. 3. Misuse case # 2: flight simulation controls and movement

4.2   Risk Assessment

The fault tree drawn for this system is based on the failure of a flight simulation, which in turn can be a result of pilot error, external factors, system error, erroneous flight data recording, failure to halt simulation, and loss of controls during simulation. This is depicted by misuse cases shown in Figs. 2 and 3. The fault tree shown in Fig. 4 represents the risks analyzed during this step of the risk mitigation strategy. The frequencies highlighted in the FTA were based on existing literature and reports on security and privacy violation incidents from past several years. The SIL levels used during the risk assessment step are recommended by DO 178-C [3] and were assigned

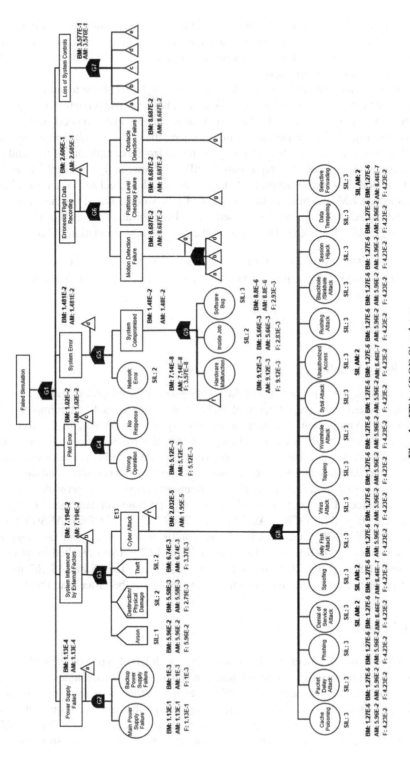

Fig. 4. FTA_6DOF Simulator

against each risk identified based on sub-system's mode of operations (continuous and higher demand rates, and lower demand rates). The resultant risk values obtained against each identified risk in base events are a product of their probability of occurrence and the assigned SIL level.

The risk values calculated for top-level events were based on the identification of all the minimal cut sets followed by the application of bottom-up approach to evaluate fault tree while using the values of likelihood of a particular base event to occur and their SIL levels (see Fig. 4). The repeated set of risks identified in the fault tree are grouped together as 'a', 'b', 'c', 'd', 'e', 'f' and 'g' to minimize the complexity of the fault tree, whereas the symbols SIL AM represents the SIL level after mitigation, and F represents the likelihood of a risk to occur. FTA also represents the calculated risk values before mitigation as BM and after mitigation as AM risk values. Based on the final set of minimal cut sets obtained and processed using Boolean algebra (see Eq. (1)). A risk value of 7.139E−1 was calculated against the risk of flight simulation failure.

$$G1 = (E8 \times E9) + E10 + E11 + E12 + E16 + E21 + E22 + E23 + E24 + E25 + E26 + E27 + E28 + E29 + E30 + E31 + E32 + E33 + E34 + E35 + E36 + E37 + E38 + E39 + E14 + E15 + E16$$

$$(1)$$

4.3   Risk Mitigation

During this step, the suggested hybrid cryptosystem was integrated with the original software system under the influence of MiTM and spoofing attack. After implementing the hybrid cryptosystem, the data sniffed out by the MiTM was in the form of a ciphertext rather than the actual data being transmitted between the various components of the software system. Unlike most cryptosystems, our hybrid cryptosystem worked faster due to the elimination of key management process in encryption and decryption of data, taking place at the nodes of the system connected over a network. The comparison conducted between some of the recently proposed or suggested encryption schemes such as AES(GCM), AES(ECB), AES(CCM), AES(CBC), AES(ECB) and ECC, and AES(ECB) and SHA3(256) as a mitigation measure and our suggested mitigation measure (RSA combined with 3DES and CMCIAA) resulted in an execution time of 566.67 ms, 26.33 ms, 424.33 ms, 25.33 ms, 123.75 ms, 181.67 ms, and 15.67 ms respectively, proving our mitigation measure to be faster than them.

5   Discussion

The software development standards proposed for different types of vehicular cyber-physical systems mainly recommend quantitative risk analysis techniques. On the other hand, when developing cyber-physical systems in other domains, such as in the medical domain, the defined standards recommend the use of a qualitative risk assessment technique. During the risk identification step, not all the risks could be identified using misuse cases but became visible through fault tree constructed during the risk assessment step. The fault tree generated during step # 3, i.e., risk assessment was mainly mapped to the risks identified during the risk identification step. The fault tree analysis recommended by DO 178-C for risk analysis gave us a detailed overview of the possible risks to be considered during the development of 6 DoF simulator. The abstract level risks identified via misuse cases were further analyzed for the identification of their possible root causes and risk value calculation. According to the risk assessment conducted, cyber-attack was the most repeated base event leading to a possible risk linked to and identified in the fault tree. The risk value of cyber-attack alone in the entire fault tree is 2.37E−4, i.e., 1/4th of the total risk value calculated for simulation failure. The overall risk value calculated before mitigation was 0.170 while the risk value calculated after mitigation was 0.168. The implementation of the hybrid cryptosystem from step # 4, i.e., risk mitigation integrated with the software system, reduced the severity of the cyber–attack from SIL 3 to Severity of SIL 2 thus reducing the overall risk value of the external factors, and system errors affecting the system's operation, and other risks linked to cyber-attacks.

The minute difference between the risk values calculated before and after mitigation was due to the fact that we focused only on a sub-set of the risks identified during the initial set of steps. In order to further reduce the overall risk of the system failure there is a need to implement other mitigation measures for the entire architecture of CPS regardless of the application domain.

6 Conclusion

The complex nature of CPS due to the requirement of constant connectivity among its components and limited resources makes it difficult to implement traditional IT security requirements. In the meantime, implementing CPS specific security requirements into its design alone is still not enough. In this article, we applied steps to identify analyze and mitigate risks in a case study specific to avionics. The results obtained are a proof of the fact that to ensure users trust into CPS, integration of risk management processes with the software development lifecycle is of utmost importance. The mitigation measure applied not only reduced the severity of the risks identified and analyzed but was also relatively faster in execution speed as compared to some of the most recently suggested and used cryptosystems.

References

1. Santini, R., Panzieri, S.: A graph-based evidence theory for assessing risk. In: 18th International Conference, Information Fusion, pp. 1467–1474 (2015)
2. Smith, D., Simpson, K.: Functional Safety: A Straightforward Guide to Applying IEC 61508 and Related Standards, 2nd edn. Elsevier Butterwirth-Heinemann, Oxford (2004)
3. Rierson, L.: Developing Safety-Critical Software: A Practical Guide for Aviation Software and DO-178C Compliance, 1st edn. CRC Press, Boca Raton (2013)
4. Hird, J., Hawley, M., Machin, C.: Air traffic management security research in SESAR. In: Proceedings - 11th International Conference on Availability, Reliability and Security (ARES), pp. 486–492 (2016)

5. Gong, L., Zhang, L., Zhang, W., Li, X., Wang, X., Pan, W.: The application of data encryption technology in computer network communication security. In: American Institute of Physics, vol. 1834 (2017)

6. Ab Rahman, N.H., Glisson, W.B., Yang, Y., Choo, K.-K.R.: Forensic-by-design framework for cyber-physical cloud systems. IEEE Cloud Comput. **3**(1), 50–59 (2016)

7. Peng, Y., Lu, T., Liu, J., Gao, Y., Guo, X., Xie, F.: Cyber-physical system risk assessment. In: 9th International Conference Proceedings on Intelligent Information Hiding and Multimedia Signal, pp. 442–447 (2013)

8. Humayed, A., Lin, J., Li, F., Luo, B.: Cyber-physical systems security – a survey. IEEE Internet Things J. **4**(6), 1802–1831 (2017)

9. Best, J.: "'Wake up baby': Man HACKS into 10-month-old's baby monitor to watch sleeping infant." Mirror Online, April 2014

10. Polemi, N., Papastergiou, S.: Current efforts in ports and supply chains risk assessment. In: 2015 10th International Conference for Internet Technology and Secured Transactions, ICITST 2015, pp. 349–354 (2015)

11. Wu, G., Sun, J., Chen, J.: A survey on the security of cyber-physical systems. Control Theory Technol. **14**(1), 2–10 (2016)

12. Manshaei, M.H., Zhu, Q., Alpcan, T., Bacşar, T., Hubaux, J.-P.: Game theory meets network security and privacy. ACM Comput. Surv. **45**(3), 1–39 (2013)

13. Cárenas, A.A., Amin, S., Sinopoli, B., Giani, A., Perrig, A., Sastry, S.: Challenges for securing cyber physical systems. In: Workshop on Future Directions in Cyber-Physical Systems Security (2009)

14. Yoneda, S., Tanimoto, S., Konosu, T.: Risk assessment in cyber-physical system in office environment. In: 18th International Conference on Network-Based Information Systems, pp. 412–417 (2015)

15. Axelrod, C.W.: Managing the risks of cyber-physical systems. In: 2013 IEEE Long Island Systems, Applications and Technology Conference (LISAT), pp. 1–6 (2013)

16. Kim, Y., Kolesnikov, V., Thottan, M.: Resilient end-to-end message protection for cyber-physical system communications. IEEE Trans. Smart Grid **9**(4), 2478–2487 (2016)

17. Rajbhandari, L., Snekkenes, E.A.: Mapping between classical risk management and game theoretical approaches. In: De Decker, B., Lapon, J., Naessens, V., Uhl, A. (eds.) CMS 2011. LNCS, vol. 7025, pp. 147–154. Springer, Heidelberg (2011). https://doi.org/10.1007/978-3-642-24712-5_12

18. Zhou, L., Guo, H., Li, D., Zhou, J., Wong, J.: A scheme for lightweight SCADA packet authentication. In: 23rd Asia-Pacific Conference on Communications (APCC) (2017)

19. Karati, A., Amin, R., Islam, S.K.H., Choo, K.R.: Provably secure and lightweight identity-based authenticated data sharing protocol for cyber-physical cloud environment. IEEE Trans. Cloud Comput. **7161**(c), 1–14 (2018)

20. Fovino, I.N.: SCADA system cyber security. In: Markantonakis, K., Mayes, K. (eds.) Secure Smart Embedded Devices, Platforms and Applications, pp. 451–471. Springer, New York (2014). https://doi.org/10.1007/978-1-4614-7915-4_20

21. Biro, M., Mashkoor, A., Sametinger, J., Seker, R.: Software safety and security risk mitigation in cyber-physical systems. IEEE Softw. **35**(1), 24–29 (2017)

22. Fletcher, K.K., Liu, X.: Security requirements analysis, specification, prioritization and policy development in cyber-physical systems. In: 2011 5th International Conference on Secure Software Integration and Reliability Improvement - Companion, SSIRI-C 2011, pp. 106–113 (2011)

Machine Learning and Knowledge Graphs

Analyzing Trending Technological Areas of Patents

Mustafa Sofean$^{(\boxtimes)}$, Hidir Aras, and Ahmad Alrifai

FIZ Karlsruhe,
Hermann-von-Helmholtz-Platz 1, 76344 Eggenstein-Leopoldshafen, Germany
{mustafa.sofean,hidir.aras,ahmad.alrifai}@fiz-karlsruhe.de
http://www.fiz-karlsruhe.de

Abstract. Analyzing technological areas of inventions in patent domain is an important stage to discover relationships and trends for decision making. The International Patent Classification (IPC) is used for classifying the patents according to their technological areas. However, these classifications are quite inconsistent in various aspects because of the complexity and they may not be available for all areas of technology specially the emerging areas. This work introduces methods that applied on unstructured patents texts for detecting accurate technological areas to which the invention relates, and identifies semantically meaningful communities/topics for a large collection of patent documents. A hybrid text mining techniques with scalable analytics service that involves natural language processing which built on top of big-data architecture are used to extract the significant technical areas. Community detection approach is applied for efficiently identifying communities/topics by clustering the network graph of technological areas of inventions. A comparison to the standard LDA clustering is presented. Finally, regression analysis methods are applied in order to discover the interesting trends.

Keywords: Patent analysis · Community detection · Topic modeling

1 Introduction

Patent documents contain a lot of important valuable knowledge, which can save time for new product development, increase success chance for market, and reduce potential patent infringement. But they are large in quantity, lengthy in space and rich in technique terminology such that they are difficult for human to sift through. A patent document often contains dozens of fields that can be grouped into two categories: First; structured fields (metadata), which are uniform in semantics and format such as patent number, inventors, and citations, filing date, issued date, and assignees. Second; textual fields, which consist title, abstract, claims, and the detailed description of the invention. The latter includes the technical field, background, summary, embodiments, and the description of figures and drawings of the invention. Patent classification schemes such as the

© Springer Nature Switzerland AG 2019
G. Anderst-Kotsis et al. (Eds.): DEXA 2019 Workshops, CCIS 1062, pp. 141–146, 2019.
https://doi.org/10.1007/978-3-030-27684-3_18

International Patent Classification (IPC) are assigned by experts to classify the patent documents according to their technologies. It is mainly based on a taxonomy, which is a hierarchical structure consisting of sections, classes, subclasses, groups and subgroups, and contains approximately 72,000 categories. Based on IPC codes, many studies have been done to obtain the features or trends of technology development by using patent search and international patent classification (IPC) [1]. However, these classifications have proved to be quite inconsistent in various aspects because of the complexity and they may not be available for all areas of technologies specially the emerging technological areas as we have observed that most of trending areas in computer science domain such as "Deep Learning", "Big Data Cloud", "Blockchain", and "Internet of Things (IoT)" are missing in description of patent classification schema, but they clearly appear in texts of the technical field segment. The technical field section in patent is a brief paragraphs, typically one to two sentences long, describing the accurate technological areas to which the invention relates, and appears in detailed description part of patent texts. The clue in this paper is to use the technological information in texts of technical field segment instead of the whole texts of patent in order to discover trending technical areas for an specific topic, as well as saving texts of technical field segment in separated field in a search index enables the interested actors to easily extract all patent documents that are related to specific technical areas. Moreover, analyzing texts of technical field segment will be benefit for the patent information professionals to understand deeply the accurate technical areas which the inventions are related to, discover trends and emerging technologies, and effective for the patent text information retrieval specially to increase the Recall. Various methods involving machine learning and text mining have been proposed in order to extract value from patent information [2,6]. Topic modeling has been applied on patent texts [3,4], and [5]. It has been applied on datasets that contained many variant topics; however, discovering topics for texts of technical field of inventions by standard topic modeling like LDA is suffering from topics overlap when analyzing a patent dataset related to an accurate topic. The methods in this work do not suffer from topics overlap, automatically determine the appropriate number of topics, via the adaptation of community detection algorithms on a network graph where vertices are technological areas of the inventions and edges are the relations between them, and discover the most important trending topics. The remaining of this paper is organized as follows. Section 2 provides our methods and analysis. Section 3 describes our experiments and results. Finally, Sect. 4 gives the conclusion and outlines the future research.

2   Methods and Analysis

The detailed description texts of the invention include many different sections, namely the technical field, background, summary, embodiments, and the description of figures and drawings of the invention. The text of technical field is short paragraphs, with three or four sentences or could be more, which displays the technological areas of the described invention. It generally mirrors the wording of the beginning of the independent claim. The aim of technical field section is to guide patent information professionals searching for related patents and understand in depth the accurate technical areas which the inventions are related. The text of technical field segment is also mirror the hierarchies expressed in more patent classification codes. For example, this text *"The invention belongs to the field of communication network, in particular relates to a domain name of the biometrics-based authentication system and method."* belongs to a new authentication method for the area communication network.

To identify the text of technical field section in patent texts, our significant and scalable segmentation methods are developed for structuring the patent texts into pre-defined sections [7]. The description of the patent text is segmented into semantic sections including technical field by using a hybrid text mining techniques such as machine learning, rule-based algorithm, and heuristics (learning-by-problem-solving pattern). Particularly, our segmentation process was used to classify each patent paragraph in description text of patent into one of several pre-defined semantic sections. The methods start with extracting a description text from the patent document, and check if the text is structured by headlines. Then the text is divided into paragraphs, a pre-processing step will take place to remove undesired tokens and apply stemming, a rule-based algorithm is used to identify the headers, and a machine learning model is used to predict a right segment for each header. For the patent that do not have complete structured segments or do not have any heading at all, heuristic methods are used. The final step is, identify boundaries of each segment and their related text content. The performance of our segmentation methods achieved up to 94% of accuracy. Sophisticated Natural Language Processing processes (NLP) are performed to automatically extract the most significant technological areas from texts of technical field segment. In particular, a large-scale, efficient, and distributed NLP pipeline is built to extract the significant noun terms and phrases from the texts of technical fields.

Patent technological areas clustering task is described as follows: given a set of technological areas, cluster them into groups so that the similar areas are in the same cluster. We used a clustering-based topic modelling algorithm that uses community detection approaches for topic modelling, based on the network graph of texts of technical field segment in patents. Then, a co-occurrence graph for technical areas is created by generating a co-occurrence matrix to generally describe how technical areas occur together that in turn capture the semantic relationships between them, since each technical area is treated as a topic (node) and the relationship between two areas (edge) is simply based on co-occurrence statistics which can indirectly represent the semantic information of related areas, and should be above a certain threshold. The weight of edges is created by using statistical measure of occurring two co-areas in a collection of patent documents. SLM algorithm [8] is used for modularity-based community detection in large networks, and it was shown to be one of the best performing algorithms in a comprehensive survey of community detection algorithms. The Maximum Likelihood is used to assign a patent document to a community/topic that has the highest co-occurrence of terms of technical areas in both the patent document and community/topic, where the term in the community/topic is weighted based on its co-occurrence to other terms.

Table 1. List of communities/topics.

Label	Detailed content
Community 1	Public key, digital signature, private key, electronic signature, digital certificate, elliptic curve, electronic transaction, common key, elliptical curve, unique identifier, hash function, public key infrastructure, encrypted message, symmetric key, shared secret, certification authority, public key certificate, public key encryption, public key cryptography, electronic commerce, certificate authority, key generation, modular exponentiation, signature generation, symmetric encryption, prime number, public key infrastructure PKI
Community 2	User authentication, authentication method, authentication system, biometric data, authentication server, biometric authentication, biometric information, personal identification, authentication device, mutual authentication, biometric identification, authentication process, personal identification number, user identification, user authentication method, authentication token
Community 3	Quantum key, quantum key distribution, quantum cryptography
Community 4	DRM system, digital content, digital management, digital management DRM, digital media, digital protection

Finally, regression analysis methods are used to estimate the trends from all communities or topics that are previously extracted, and the trend line for each topic is drawn. In particular, for each extracted community/topic, the number of related patents is calculated and distributed over years, trend curve is drawn and a moving average is used to smooth that curve, coefficient of determination are calculated, and then only upward trends are selected to be the hotspot topics.

3   Experiments and Results

This work analyzes the patent dataset that is related to the "Information Security and Cryptography" domain. From the databases Patent Cooperation Treaty and European Patents, 34340 patents are extracted. We applied our MapReduce-based segmentation tool to extract the text of technical field section from each patent in the dataset [7,9], Spark framework is used to perform our NLP tasks in order to extract the most important technical areas of inventions, then we applied Spark Latent Dirichlet Allocation (Spark-LDA) to extract the list of topics. It was found that some terms of technical areas are shared between many different topics. For instance, the term "authentication" is shared between five different topics. However, this usually makes it much harder to find more accurate areas

Table 2. Performance of methods of technological area extraction.

Data	P@50	P@100	P@150	P@200	MAP
Patent dataset	92%	90%	88%	87%	89%

such as "fault attack", "biometric authentication", and "access control" that are clearly appearing in our dataset. Also LDA has been experimented on sub topics in "Information Security" domain such as cryptography, and we found that LDA is suffering much more from topics overlap in case of analyzing texts of technical fields of inventions.

To show the results of our methods in this work, the occurrence number is calculated for each technical area that is extracted by our NLP processes. A scalable Spark-based service is developed to create the network graph for most frequent areas via calculating a co-occurrence matrix. The co-occurrence matrix is computed by counting how two or more technical areas occur together in our corpus. For community detection, the open source codes of Smart Local Moving Algorithm is integrated into our Spark application in order to cluster network graph into communities/topics, and the community with only one or two technical areas are ignored.

Table 1 only presents four communities/topics from all topics that are discovered by our methods in this work. As shown in the table, all technical areas that related to the topic "public key infrastructure" are grouped in the community 1, community 2 contains all terms that are related to a topic "authentication", community 3 is related to "quantum cryptography", and community 4 is related to "digital rights management". Moreover, more promised technical areas appeared in our results such as "security management", "access control", "data integrity", "personal information (privacy)", as well as certain more specific areas have appeared in the results which represent the subject matter of inventions like "one-time password", "fault attack", "biometric authentication", and "single sign-on". In comparison with standard topic modeling like LDA, our methods do not suffer from topics overlap, and do not require the setting of numerous parameters. Moreover, we found that using community detection for topic modeling works well with real applications of patent analysis without sharing terms between communities/topics. For instance, the terms in the community 3 and 4 in the Table 1 are related to "quantum cryptography" and "DRM" respectively without any term interference from other communities.

For the evolution of patent technological areas extraction, we independently requested three data scientists to provide human judgement on the top (200) returned terms. The judgments about relevant (Like) or irrelevant (DisLike). If there are more than two annotators saying that an extracted technological area is irrelevant, we remove it from the returned list. We used the Precision to measure the ability of our methods to reject any non-relevant technical area in the retrieved set. Therefore, the evaluation is conducted in terms of P@N (Precision for top N results), mean average precision (MAP) [10]. The performance results of the evaluation are presented in the Table 2. The methods in this work were developed based on a combination of Apache Hadoop and Apache Spark frameworks, and are integrated into scalable patent mining and analytics framework that is built on top of big-data architecture and a scientific workflow system for allowing the user to efficiently annotate, analyze and interact with patent data on a large-scale via visual interaction [9].

4   Conclusion

In this paper we have described our methods for detecting texts of technical field in unstructured patents texts, and extracted technological areas of inventions. We presented a large-scale graph-based technical area clustering methods that utilize community detection approaches for aggregating technical areas into communities/topics based on a network graph in order to discover trends for a large collection of patent documents. A comparison to the standard LDA clustering has been presented, and the results showed that our methods do not share common terms and perform well with real patent applications. Our future directions of research will continue to analyze technical areas in patent domain by using deep learning graph, and improve the semantic between technological areas of patents by ontological knowledge and linked data.

References

1. Yan, B., Luo, J.: Measuring technological distance for patent mapping. J. Assoc. Inf. Sci. Technol. 68, 423–437. https://doi.org/10.1002/asi.23664
2. Abbas, A., Zhang, L., Khan, S.U.: A literature review on the state-of-the-art in patent analysis. World Patent Inf. **37**, 3–13 (2015)
3. Ankam, S., Dou, W., Strumsky, D., Zadrozny, W.: Exploring emerging technologies using patent data and patent classification. In: CHI 2012, Austin, Texas, USA. ACM (2012)
4. Chen, H., Zhang, Y., Zhang, G., Lu, J.: Modeling technological topic changes in patent claims. In: Proceedings of PIC MET 2015, Portland, OR, USA (2015)
5. Tang, J., Wang, B., Yang, Y., Hu, P., Usadi, A.K.: PatentMiner: topic-driven patent analysis and mining. In: KDD 2012, Beijing, China. ACM (2012)
6. Trippe, A.: Guidelines for Preparing Patent Landscape Reports. Patinformatics, LLC, With contributions from WIPO Secretariat (2015)
7. Sofean, M.: Automatic segmentation of big data of patent texts. In: Bellatreche, L., Chakravarthy, S. (eds.) DaWaK 2017. LNCS, vol. 10440, pp. 343–351. Springer, Cham (2017). https://doi.org/10.1007/978-3-319-64283-3_25
8. Waltman, L., van Eck, N.J.: A smart local moving algorithm for large-scale modularity-based community detection. Eur. Phys. J. Springer (2013)
9. Sofean, M., Aras, H., Alrifai, A.: A workflow-based large-scale patent mining and analytics framework. In: Damaševičius, R., Vasiljevienė, G. (eds.) ICIST 2018. CCIS, vol. 920, pp. 210–223. Springer, Cham (2018). https://doi.org/10.1007/978-3-319-99972-2_17
10. Buckley, C., Voorhees, E.M.: Retrieval evaluation with incomplete information. In: SIGIR 2004, Sheeld, South Yorkshire, UK, pp. 25–32. ACM (2004)

Emotion Analysis Using Heart Rate Data

Luis Alberto Barradas Chacon, Artem Fedoskin[(⊠)],
Ekaterina Shcheglakova, Sutthida Neamsup, and Ahmed Rashed

Information Systems and Machine Learning Lab, University of Hildesheim,
Hildesheim, Germany
{barradas, fedoskin, shchegla, neamsup}@uni-hildesheim.de,
ahmedrashed@ismll.de

Abstract. This paper describes the attempt to classify human emotions without including Electroencephalography (EEG) signal, using the DEAP dataset (A Database for Emotion Analysis using Physiological Signals). Basing our research on the original paper (Koelstra et al. 2012), we claim that emotions (Valence and Arousal scores) can be classified with only one pulse detecting sensor with a comparable result to the classification based on the EEG signal. Moreover, we propose the method to classify emotions avoiding any sensors by extracting the pulse of the person from the video based on the head movements. Using Lucas-Kanade algorithm for optical flow (Balakrishnan et al. 2013), we extract movement signals, filter them, extract principal components, recreate heart rate (HR) signal and then use in the emotion classification. The first part of the project was conducted on the dataset containing 32 participants' heart rate values (1280 activation cases), while the second part was based on the frontal videos of the participants (874 videos). Results support the idea of one-sensor emotion classification and deny the possibility of zero-sensor classification with the proposed method.

Keywords: Emotion classification · DEAP dataset · Heart rate estimation · Biosignals

1 Introduction

Emotional assessment is a common problem from the field of psychology where the emotional state of a person is to be deduced. For this, there are two common approaches: physiological assessment, and self-report. The latter approach is done through a verbal report, where the subject responds to specific questions about their emotional state. This approach considers the subject's experience as a subjective view, and relies on his ability to verbalize his internal states. A more objective way to assess emotions is through biosignal inference, as mentioned as physiological assessment, where the emotional state of the subject is deduced from their physiological responses to stimuli. This method relies on normative physiological responses through different populations. From this approach the common biosignals inferring emotional states are:

- Plethysmography or Blood volume pulse (BVP): an optical device measures the volume of blood flowing through an extremity, and records changes of each beat of

G. Anderst-Kotsis et al. (Eds.): DEXA 2019 Workshops, CCIS 1062, pp. 147–154, 2019.
https://doi.org/10.1007/978-3-030-27684-3_19

the heart. With this, the Heart Rate (HR), and other heart-related variables, such as the beat per minute (BPM) and the Heart Rate Variability (HRV) can be inferred.

- Electroencephalography (EEG): Measuring the electrical activity that emanates from the skull can be the most direct way to measure cognitive and emotional changes, but EEG signals are complex and require much processing.

Like with any assessment task, a model of the desired states must be selected. Emotional states vary greatly: Ekman's emotional model (Ekman 1999) presents six human emotions shared across all cultures through facial expressions. Plutchik's wheel of emotion presents eight common responses to stimuli. One of the most simple models: the affect-activation model is commonly used in text analysis. It allows representing an emotional state in a cartesian plane embedding, where one axis represents the level of activation, and the other the valence of such stimuli: how pleasurable it is. This last model was used in this experiment, due to its low-dimensionality, and dichotomy.

In this paper, we try to improve over the results from (Koelstra et al. 2012), in which authors used EEG data from DEAP dataset to predict low or high Valence/Arousal scores.

Section 2 presents related work. Section 3 presents the methodology. Section 4 presents the results. Finally, Sect. 5 concludes our paper.

2 Related Work

One way to study human behavior and emotion recognition in a quantitative way, is to measure various biometric parameters associated with an emotional state or a stress response to the given stimulus. After measuring biosignals such as EEG, heart and breath rates, skin conductance, temperature, etc. those measures can be used for analysis with Machine Learning algorithms, which can help to find hidden insights in data and make it applicable to real-life situations.

In 2012 the group of researchers introduced the DEAP (Koelstra et al. 2012) dataset (A Database for Emotion Analysis using Physiological Signals), which became the emotional classification benchmark. Using EEG signals for the Gaussian Naive Bayes classifier, authors were able to classify arousal and valence ratings with the following accuracy: 62% and 57.6% accordingly. After the release of this work, researchers tried to use different algorithms for emotional recognition to improve the results of the DEAP dataset paper. Bayesian Classification methods for classifying Valence and Arousal from the DEAP dataset into two classes – high and low – were introduced in 2012 and achieved 66.6% and 66.4% accuracy correspondingly (Chung and Yoon 2012). In their work, (Hosseini et al. 2010) raises an electroencephalogram (EEG) method in measuring the Biosignal in order to classify emotional stress. He introduces the visual images-based acquisition protocol for recording the EEG together with psychophysiological signals under two categories of emotional stress states, then does the classification with SVM classifier algorithm. Therefore, the results in this paper suggest that using Support Vector Machine classifier also gives better accuracy to classify categories of states compare among to previous research out in the field. The use of Support Vector

Machines as a classifier for the EEG data for the real-time emotion recognition using Higuchi Fractal Dimension Spectrum was introduced in 2014 and produced 53.7% accuracy for classification of 8 emotions (Liu and Sourina 2014). In 2015 the effect of the window size on the emotion classification on EEG data using wavelet entropy and SVMs was studied (Candra et al. 2015). A too short window will not allow the information about emotions to be extracted fully, while the too wide window will lead to the information overload. As a result, the following accuracies were received after the experiments for Valence and Arousal scales: 65.13% with 312 s window and 65.33% with 310 s respectively. In one of the most recent works the Deep and Convolutional Neural Networks approaches were tested and brought the best for the current moment results for the emotional classification: 81.4% for Valence and 73.4% for Arousal using CNN (Tripathi et al. 2017).

However, there also exists the problem of collecting such type of data like EEG for the following analysis. To do so you need to cover a person with different sensors, staying still without any movements, everything should be under strict control. To prevent this, we claim that there is no need in those sensors and with only a pulse detector you can classify emotions like with the EEG signal. Moreover, we suggest the method to classify emotions without any sensor based on the video of the person.

Three signals (heart rate, breath rate, and skin conductance) are suggested to be linked with physiological activation (Choi 2010; Lazarus 1963) along with the heart rate variability (HRV), which is not the pulse, but the heart rate over time, which is also linked to stress (Mohan 2016). The work of (Balakrishnan et al. 2013), on which our zero-sensor classification approach is based, presents the novel approach of cardiac monitoring without actual contact-measuring means. The research suggests to use the video, capture the motion of the movement features of the blood flow from the heart to the head frequencies and extract the features by using principal component analysis (PCA) to decompose the trajectories into a set of independent source signals.

3 Methodology

3.1 Dataset Description

In this paper, the DEAP dataset is used. It can be separated into two parts: the recorded signals from all devices and self-assessment report of the subjects. Firstly, the electroencephalogram (EEG) and peripheral physiological signals (Pulse, Breath, GSR, Temperature, BVP) were recorded for the 32 participants, which were watching 40 one-minute music videos with most intensive emotional parts (in total 1280 emotional activation cases). For 22 participants also the frontal videos were recorded, on which our zero-sensor classification method is built. Secondly, after each video, participants rated the videos based on arousal, valence and dominance scales. Each video has a subjective rating of the levels 0–9 of arousal and valence from each participant, which is used as the ground truth for classification in the current paper. To make it comparable to the original paper, the scale was also separated into two classes: high (5–9) and low (0–4) for both valence and arousal scales.

Our primary goal was to predict excitement in terms of arousal and valence scales of the participant. Our first approach was to use HR data produced by the sensor (given in the dataset) and the second one was to calculate HR of a participant using the facial recording. Due to the number of participants and frontal recordings provided by the dataset, 1280 samples for the first approach and 874 activation cases for the second one were used.

3.2   One-Sensor Classification

We used a heart rate data that was collected as a pulse rate by using a plethysmography device on the subject's thumb. It was later filtered with Butterworth filter (SciPy library) to remove noises. To accomplish that, Scikit-Learn library (Pedregosa et al. 2011) was used to retrieve peak-to-peak measurement. Afterward, we extracted Beats Per Minute and Heart Rate Variability from filtered heart rate data that are used as features in our models. Figure 1 shows the cleaned pulse signal and calculated peaks of heart rate signal. 70% of the data was used for training and the remaining 30% served as a test set.

We used several models to test our hypothesis, namely - Support Vector Machines, Recurrent Neural Network, and Feedforward Neural Network. The choice of the models was made due to the low-dimensional nature of data, thus the Convolutional Neural Network was not used because our main goal was to prove that emotion classification can be done not only with EEG but also with basic sensors that produce such data as HRV and BPM. Finally, RNN was chosen due to the time-series nature of pulse data.

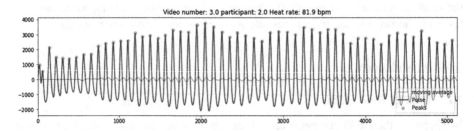

Fig. 1. The plot of HR data sample (from video number 3 and participant number 2)

Our models used several features that were extracted from the Heart Rate data:

- Beats Per Minute (BPM) - the number of heart beats per minute.
- Heart Rate Variability (HRV) - the difference between two adjacent peaks measured in milliseconds.
- Pulse was calculated as BPM per every second of the trial.

Our first attempt to classify emotional state based on the heart rate was done with SVM as a classification model. This is a well-known model that is famous for its robustness and simplicity, thus we used this model only with 2 features - HRV and BPM.

The second model we used was a Feedforward Neural Network that had 4 hidden layers each followed by ReLU activation function. This model was designed to handle complex relations as apart from HRV and BPM we used Pulse as one of the features. This method proved to be the best among all the models tested.

Our last attempt was done using Recurrent Neural Network that had 4 hidden layers with 100 hidden units in each layer. We used only one feature, namely BPM for 45 s of each trial.

Finally, all experiments with Neural Networks were done with PyTorch and GeForce GTX 1050Ti was used for training the models.

3.3   Zero-Sensor Classification

In zero-sensor approach, our method was to classify participants emotions based solely on the facial video. In order to do that we used HR extraction technique (Balakrishnan et al. 2013) that (1) uses the Lucas-Kanade (LK) algorithm to detect optical flow, (2) extracts movement in the video as signals, (3) filters them accordingly to the expected frequencies of a heart rate, (4) is separated into its principal components, (5) to finally be recreated as HR signals.

To accelerate the processing of the LK image flow algorithm, interest points were first selected inside the area of the image that contained the face of the participant. This was done through a fast-corners feature point detection algorithm. These interest points were later used as the input to the LK algorithm with a 15×15 pixel window, meaning that the flow of the video was only detected in the nearest region. The result was time signals that contained the optical flow inside that image window along with the video. These signals were processed through a temporal filter that removed frequencies that do not correspond to heart rate and later decomposed through a Principal Component Analysis (PCA). This last decomposition separates the signals that most similarly resemble HR, and those that contain more noise. The components most similar to a heart rate were then selected to recreate a heart rate signal. This selection was done manually for every subject, through visual inspection.

The reconstruction of a heart rate signal takes about 12 min for every minute of video, making it unlikely to be used in real time (Fig. 2).

One of the problems we encountered was the choice of principal components; since every interest point was eventually analyzed as a time signal, we could obtain a component for every interest point. We could not select the most similar one to a heartbeat automatically. As a result, the recreated signal, although it had the same or very similar frequency as a heart rate, rarely correlated with the original one in the time domain. A workaround for this problem could be to use the real HR signal to find the component with the highest correlation to the BVP, but this would render the method meaningless, since the real signal can be used instead. Thus, we decided to try classifying emotions based on the raw reconstructed signal. This signal was used as a single feature for Recurrent Neural Network model, which, as it is shown in the results section, did not perform well.

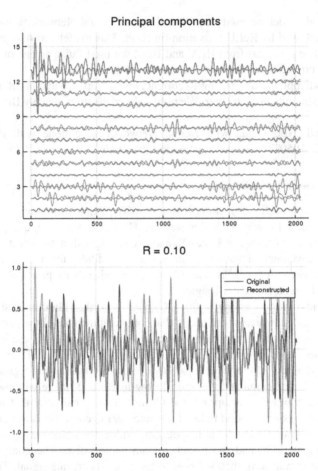

Fig. 2. Up - the plot of all principal components found in the facial video. Down - the original signal and the reconstructed one. R - correlation coefficient

4   Results

DEAP dataset provides Valence and Arousal values for each participant on the 1 to 9 scale. In order to simplify our trials and follow the same procedure as baseline paper did, we converted Valence and Arousal values to binary values with Valence/Arousal of 1–5 representing low values and the remaining ones are labeled as high values (Table 1).

Table 1. Class balance

Class balance	Valence	Arousal
High	43%	42%
Low	57%	58%

As it is evident from the results in Table 2 our model performed better than (Koelstra et al. 2012) proving our hypothesis that one can classify Valence and Arousal purely using heart rate data available and get results better than the ones obtained with EEG data.

The combination of pulse signals and HRV with BPM made FNN perform the best in the classification of Valence. There is a 3% improvement over the baseline case given by Gaussian Naive Bayes, which used EEG signals. This shows the capability of using pulse rate in measuring the valence and arousal. Moreover, for Arousal activation, Support Vector Machine classification method produced the best results. Overall, those two models performed better than the Recurrent Neural Network classification model.

RNN trained with a pulse did not perform the worst with average results, showing that Pulse is a significant feature, but it can perform much better in conjunction with HRV and BPM, which is proven by results of FNN.

Finally, RNN trained on signals obtained within zero-sensor approach did not perform as good as expected because of the problems within the method of HR extraction.

Table 2. Accuracy of models (red color denotes best results of our models)

Classification Model	Valence	Arousal
Gaussian Naive Bayes (EEG) [Koelstra et al, 2012]	57.6%	62.0%
SVM (HRV and BPM)*	55.5%	61.6%
RNN (Pulse)*	59.6%	58.5%
FNN (HRV, BPM, Pulse)*	60.6%	57.0%
RNN (Signal from Video)**	57.0%	58.0%

* - One-sensor approach (DEAP dataset), ** - Zero-sensor approach

5   Conclusion

Our results show 3% improvement in accuracy for valence classification and comparable performance for arousal classification comparing to the baseline paper. These numbers indicate that more work should be done in the field of non-EEG based emotion analysis. However, the method that we initially proposed for zero-sensor testing did not meet our expectations as we encountered several problems during the implementation and testing stages.

Based on our results we see several directions for our future work:

1. Use color differences in the red channel of the video, that has already been tested for the Eckman model.
2. Use an infrared camera to enhance biosignals that we are using.
3. Use a CRNN to automatically learn the pulse signal from the video and predict sentiment scores in a multitasking model. This method promises to learn all the hyperparameters for previous tasks automatically and poses the best course of action for solving this problem.

References

Balakrishnan, G., Durand, F., Guttag, J.: Detecting pulse from head motions in video. In: Proceedings of the IEEE Conference on Computer Vision and Pattern Recognition, pp. 3430–3437 (2013)

Candra, H., et al.: Investigation of window size in classification of EEG-emotion signal with wavelet entropy and support vector machine. In: 2015 37th Annual International Conference of the IEEE Engineering in Medicine and Biology Society (EMBC), pp. 7250–7253. IEEE (2015)

Choi, J., Gutierrez-Osuna, R.: Estimating mental stress using a wearable cardio-respiratory sensor. In: Sensors 2010, IEEE, pp. 150–154. IEEE (2010)

Chung, S.Y., Yoon, H.J.: Affective classification using Bayesian classifier and supervised learning. In: 2012 12th International Conference on Control, Automation and Systems, pp. 1768–1771. IEEE (2012)

Ekman, P.: Basic emotions. In: Handbook of Cognition and Emotion, pp. 45–60 (1999)

Hosseini, S.A., Khalilzadeh, M.A., Changiz, S.: Emotional stress recognition system for affective computing based on bio-signals. J. Biol. Syst. 18(spec01), 101–114 (2010)

Koelstra, S., et al.: DEAP: a database for emotion analysis; using physiological signals. IEEE Trans. Affect. Comput. 3(1), 18–31 (2012)

Lazarus, R.S., Speisman, J.C., Mordkoff, A.M.: The relationship between autonomic indicators of psychological stress: heart rate and skin conductance. Psychosom. Med. 25(1), 19–30 (1963)

Liu, Y., Sourina, O.: EEG-based subject-dependent emotion recognition algorithm using fractal dimension. In: 2014 IEEE International Conference on Systems, Man, and Cybernetics (SMC), pp. 3166–3171. IEEE (2014)

Mohan, P.M., Nagarajan, V., Das, S.R.: Stress measurement from wearable photoplethysmographic sensor using heart rate variability data. In: 2016 International Conference on Communication and Signal Processing (ICCSP), pp. 1141–1144. IEEE (2016)

Pedregosa, F., et al.: Scikit-learn: machine learning in Python. J. Mach. Learn. Res. 12(Oct), 2825–2830 (2011)

Tripathi, S., Acharya, S., Sharma, R.D., Mittal, S., Bhattacharya, S.: Using deep and convolutional neural networks for accurate emotion classification on DEAP dataset. In: Twenty-Ninth IAAI Conference (2017)

Towards Explaining Deep Neural Networks Through Graph Analysis

Vitor A. C. Horta$^{(\boxtimes)}$ and Alessandra Mileo

Insight Centre for Data Analytics at Dublin City University, Dublin, Ireland
{vitor.horta,alessandra.mileo}@insight-centre.org

Abstract. Due to its potential to solve complex tasks, deep learning is being used across many different areas. The complexity of neural networks however makes it difficult to explain the whole decision process used by the model, which makes understanding deep learning models an active research topic. In this work we address this issue by extracting the knowledge acquired by trained Deep Neural Networks (DNNs) and representing this knowledge in a graph. The proposed graph encodes statistical correlations between neurons' activation values in order to expose the relationship between neurons in the hidden layers with both the input layer and output classes. Two initial experiments in image classification were conducted to evaluate whether the proposed graph can help understanding and explaining DNNs. We first show how it is possible to explore the proposed graph to find what neurons are the most important for predicting each class. Then, we use graph analysis to detect groups of classes that are more similar to each other and how these similarities affect the DNN. Finally, we use heatmaps to visualize what parts of the input layer are responsible for activating each neuron in hidden layers. The results show that by building and analysing the proposed graph it is possible to gain relevant insights of the DNN's inner workings.

Keywords: Explainable AI · Deep learning · Graph analysis

1 Introduction

In deep learning, given an extensive set of examples, Deep Neural Networks (DNNs) can be efficiently trained to solve complex problems, such as speech and language processing and image recognition. These tasks often involve high-dimensional data and, while most traditional machine learning methods (ML) fail in these cases, DNNs have been shown to achieve good performance [11].

Although this approach has proven to be powerful, there are still several challenges and open issues regarding the use of this technology. A known gap in the field is the lack of interpretability and the problem is that it is a difficult task to determine why a DNN makes a particular decision [16]. In some domains, such as precision medicine and law enforcement, it is not sufficient to know only the final outcome and how good it is, but it is also important to explain the whole

© Springer Nature Switzerland AG 2019
G. Anderst-Kotsis et al. (Eds.): DEXA 2019 Workshops, CCIS 1062, pp. 155–165, 2019.
https://doi.org/10.1007/978-3-030-27684-3_20

decision making process. Besides explaining the decisions, understanding the inner workings of a DNN could be useful to improve and extend DNN models and architectures, as well as provide useful knowledge that can be reused for transfer learning and reasoning in deep networks.

Some of the challenges involved in this task are the different and complex architectures of deep networks and the difficult to extract knowledge from neurons in hidden layers. This problem has being known and explored since the 90's, when DNNs were not yet as successful as they are now, due to the lack of available training data and computational resources. Existing approaches try to explain DNNs through rule extraction [3,5,16], feature visualization [4,14], measurement of input contribution [1,12] among others. These approaches do not generally tackle the hidden layers and some do not apply to convolutional layers. Instead, some of them use the neural network as an oracle and use its predictions to build other ML models such as decision trees, while others explain only the influence of inputs in feature space over the model outcome. These can be seen as limitations in their attempt to provide a general mechanism to extract knowledge from deep learning models and use it to explain the model itself.

In this paper a novel way to extract and represent knowledge from trained DNNs is proposed. We use activation values to find correlations between neurons at each layer in the DNN, including input layer, dense and convolutional layers and output classes. We represent this information in a graph where nodes correspond to neurons that are connected through weighted edges based on correlations in their activation values. Our hypothesis is that knowledge contained in the proposed graph is compatible with knowledge acquired by the DNN and by using graph analysis tools we can gain insights on how the model works.

In this initial work we focus on evaluating whether the proposed graph is capable of representing knowledge about how the hidden layers of the DNN work. Besides the proposed graph, three main contributions of this paper are: (i) to show how to exploit the proposed graph in order to detect which neurons are relevant to each output class; (ii) to use graph analysis for detecting groups of classes with high similarity; (iii) to visualize features in hidden layers and understand the role of neurons in both fully connected and convolutional layers.

Two initial experiments on image classification were conducted to demonstrate our approach. We show that by querying the proposed graph it is possible to find hidden layer neurons highly correlated to output classes and that, when activated in the DNN, these neurons have a high contribution to the prediction value of these classes. Also, graph analysis is performed over the proposed graph to find groups of classes that are more similar to each other. We show that classes with high number of shared neurons (overlapping nodes in the graph) are responsible for most mistakes in the DNN. This is an evidence that overlapping nodes in the proposed graph might indicate which neurons in the DNN should be considered for a fine-tuning process in order to improve the model to better distinguish the overlapping classes. In addition, edges between hidden and input layers allow us to visualize each neuron in both convolutional and dense layers through heatmaps, which helps identifying their role in the decision process.

The rest of the paper is organized as follows. Section 2 shows existing approaches to extract knowledge from neural networks. Section 3 presents the proposed methodology and explains how to generate the proposed graph. In Sect. 4 a feasibility study is conducted in two experiments on image classification. Section 5 provides our conclusions and what can be done in future works.

2   Related Works

Explaining neural networks is a difficult task that attracts researchers' attention since the 90's and many studies attempt to address this problem from different perspectives. One common approach is to measure the contribution of each input to the outputs of the neural network [1], which helps understanding the model's decision process. This approach can also be extended for measuring the contribution of neurons in the hidden layer [12], but one limitation in the existing works is that they can not be applied to convolutional layers.

Another way to understand the underlying mechanism of neural networks is through rule extraction. In this case some works adopt a decompositional strategy, which extract rules by examining activation and weights in the neural networks [3,8,9]. Another rule extraction strategy is the pedagogical, which uses the decisions made by DNNs to extract rules without exploring the model architecture [5]. It is also possible to use a hybrid of these two approaches, which is the eclectic strategy [13]. The problem is that methods that uses decompositional strategies can not be directly applied to convolutional layers. On other hand, pure pedagogical methods can be applied to any neural network because they do not rely on the architecture. However, they do not explain the inner workings of hidden layers since they keep the DNN as a black box.

A third approach is to use visualization techniques [4,14]. Visualizations can be used to understand which parts of the input are relevant for the model's predictions. It is also possible to understand the role of filters in convolutional layers by visualizing them. Still, performing these visual analyses might require a lot of human intervention, which limits their scalability.

The graph representation proposed in this paper differs from existing works in two main aspects. Firstly, our proposal can represent both convolutional and fully connected layers in any depth of DNNs. Second, it allows the use of automatic methods such as graph analysis tools to discover interesting patterns about the internal workings of DNNs. In this sense, a recent work [6] that proposes a graph representation for embedding vectors is the most similar to our proposal. One difference is that because our graph uses weighted edges, our approach can represent both negative and positive relationships with less nodes than their approach, which uses different nodes for positive and negative contributions. We also include the input layer in our representation, which allow us to visualize of neurons in the hidden layer to understand their roles in the decision process.

3   General Approach: How to Generate the Co-activation Graph

The main goal of this paper is to extract and represent knowledge from trained DNNs in order to better understand how the hidden part of the model works. We propose a graph that connects every pair of neurons of any type (fully connected or convolutional) and located in any layer of the neural network. This section presents the general idea on how to build the proposed graph and in next section we conduct preliminary experiments to assess the feasibility of our approach.

In the proposed graph nodes correspond to neurons in the DNN and weighted edges represent a statistical correlation between them based on their activation values. We call this graph as *co-activation graph*, since the relationships between its nodes represent how their activation values are correlated. The main idea of the co-activation graph is to create a relation between neurons in any depth of hidden layers to neurons in the input feature space and output classes, since the latter are more comprehensible for humans. Given a trained DNN we can generate a co-activation graph using the three steps below:

Extract Activation Values : The first step is to feed the DNN with some data samples and extract activation values for each prediction. For dense layers this process is straightforward because each neuron outputs a single activation value. Filters in convolutional layers, instead, will output multiple values since they have different activation values for each region in the input. To overcome this and make our approach work for convolutional layers, the average pooling technique is applied to extract a single value for convolutional filters. Although some spatial information is lost in this process it allows the extraction of a single activation value for each filter while keeping the dimensionality low [7].

Define and Calculate Edge Weights : After collecting the activation values for neurons in each layer, the next step is to define the strength in their relationships. We use the activation values to calculate a statistical correlation between each pair of neurons. In this work we chose to use Pearson coefficient as an initial correlation measure and thus edge weights vary in range $[-1, 1]$.

Build and Analyse the Co-activation Graph : In this third and final step the co-activation graph can be built and analysed. The nodes represent neurons of any layer in the DNN and weighted edges indicate the correlation between their activation values. We can then explore the graph structure and use graph analysis tools to understand relationships between neurons in hidden layers with the input layer and output classes.

In next section we do a preliminary evaluation of our approach to demonstrate whether the co-activation graph is a suitable way to interpret and understand the way the DNN works and the knowledge encoded in its model.

4 Preliminary Experiments and Initial Results

To evaluate whether the co-activation graph can help understanding how deep learning models work we have conducted two experiments. This section first introduces the datasets and how the experiments were conducted. Then, three analyses are performed, which are the main contributions of this work. In the first analysis we use the co-activation graph to check how relationships between nodes in hidden layers and classes in the output layer can reveal neurons that highly impact the prediction value for these classes. This is the first step to evaluate whether the co-activation graph representation is compatible with the knowledge within the trained DNNs, since it can be used to identify which neurons have high contribution over the decisions taken by the model.

In a second analysis we show how graph analysis performed over the co-activation graph can give insights about the internal workings of the corresponding DNN. For this we use a community detection algorithm to see what properties in the DNN can be revealed by analysing the community structure of co-activation graphs. The third analysis aims to discover the role of each neuron in the DNN. We use relationships between nodes in the input layer and nodes in hidden layers to produce heatmaps for neurons in both convolutional and fully connected layers. The three analyses are used to evaluate the feasibility of using co-activation graphs to extract and represent knowledge from trained DNNs.

4.1 Datasets and Experiments Setup

In our experiments we used two well known datasets: MNIST handwritten digits [10] and MNIST fashion [17]. Both datasets contain 70000 images separated in ten classes. The classes in *handwritten digits* dataset refer to digits from 0 to 9 while classes in *fashion* dataset are related to clothes. The DNN used for *handwritten digits* contains two convolutional layers and three fully connected layers and the DNN used for *fashion* dataset has three convolutional layers and two fully connected layers. These models achieved an accuracy higher than 97% and although it is possible to find more accurate models we chose to use these ones since we also want to analyse the reason behind mispredictions.

To build a co-activation graph for each DNN the three steps described in the previous section were applied. We have first fed the DNNs with data samples from the testing set in order to extract activation values for each neuron. Then, we calculated the Pearson correlations between those neurons and built a co-activation graph for each DNN. Figure 1 shows the two graphs, where yellow nodes represents neurons in the output layer (classes) and blue nodes are neurons in hidden layers. The input layer was omitted in this visualization for clarity.

4.2 Analysing the Relation Between Co-activation Graphs and DNNs

We can see from Fig. 1 that for both graphs the visualization technique placed some classes (yellow nodes) very close to each other while keeping other classes

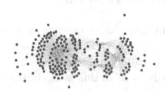 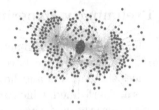

Fig. 1. Visualizations for co-activation graphs (handwritten digits left and fashion right) made on Gephi. Blue nodes are representing neurons in hidden layers and yellow nodes are representing neurons in the last layer (output classes) (Color figure online)

more distant. This might indicate that co-activation graphs have some potential to have a community structure. However, before performing such graph analysis it is important to analyse whether these graphs are being able to represent knowledge from their respective DNNs. To check this we have analysed how we can explore the structure of co-activation graphs to understand the DNNs' decision process.

By querying the graph it is possible to collect for each class a set of its most correlated neurons based on relationship weights. Since strong relationships indicate high positive correlations in the activation values, it is expected that when these highly correlated neurons are activated in the DNN, they will have a high contribution to increase the prediction value of their classes. For example, if we activate neurons most correlated to a class *sandal*, it is expected that the DNN will output *sandal* as a prediction.

To show this we extracted and manually simulated the activation of the *top k* most correlated neurons for each class. These neurons were extracted from the first fully connected hidden layer after the convolutional layers, which is not directly connected to the output and thus more difficult to interpret. We expect that when activating the top k neurons, the model will output a prediction with the respective class. Figure 2 show the results for both datasets. It is possible to see that for the *fashion* dataset when we activate the *top* 7 neurons we get the expected prediction for all the classes and for *handwritten* dataset the model predicts the expected class 80% of the times.

This result indicates that we can query the co-activation graph to find neurons with a high impact over the prediction values for each class. Therefore, it is possible to extract relevant neurons for each class from the co-activation graph even though they are in a layer located in the middle of the DNN, which is the first contribution of this paper. In a subsequent step we performed other graph analysis to see how the co-activation graph can be used to help getting a better understanding of the deep learning models.

4.3   Community Structure Analysis

Following the intuition from Fig. 1, we analysed the community structure in this graph to see if it can help detecting classes that are similar from the DNN point

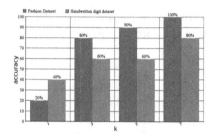

Fig. 2. Result acquired after extracting *top k* most correlated neurons for each class in the co-activation graph and activating then in the DNN. The accuracy indicates how many times the expected class was predicted.

of view. The Louvain community detection algorithm [2] was chosen because besides being a well established algorithm, this method also outputs a modularity coefficient that can be used to check how the community structure differs from random graphs. The value of modularity varies in range $[-1, 1]$ and higher values indicates that connections between nodes in same community are stronger than nodes in different ones.

We used the algorithm over the connections between nodes representing neurons in the hidden layers and nodes representing classes in the output layer. Table 1 shows the detected communities and classes contained in each of them. For the *fashion* dataset it can be noted that classes in same communities have a similar semantic meaning. We can notice this because classes like *pullover* and *coat* were put in same community while *sandals* and *sneakers* are in another one. For the *handwritten* it is not possible to conclude the same, since semantic of digits are less clear. It is important to note that for both datasets the modularity was higher than 0.4 which means that found communities are denser regions although they are not totally distinguishable.

Table 1. Classes and their communities in both datasets.

	Fashion	Handwritten digits
Community	Classes	Classes
C1	T-shirt/Top; Pullover; Coat; Shirt	0; 2; 4; 6
C2	Trouser; Dress	5; 7; 8; 9
C3	Sandal; Sneaker; Bag; Ankle Boot	1; 3
Modularity	0.413	0.45

After noticing that some classes are more similar to each other we investigated the similarities between classes and how they impact the DNNs decisions. The jaccard similarity coefficient was calculated based on the overlaps (shared nodes) between each pair of classes. Then, we collected the number of mistakes between them by counting for every pair of classes A and B how many times the model wrongly predicted A when the correct answer was B, or the other way around.

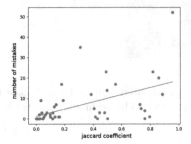 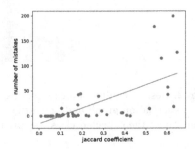

Fig. 3. Correlations of mistakes between two classes and their jaccard similarity. The Pearson correlation is 0.5042 for handwritten digits (left) and 0.6625 for fashion (right)

Figure 3 shows that there is a positive correlation between class similarity and the number of mistakes involving them. This shows that pairs of classes with many overlapping nodes in the co-activation graph tend to cause a high number of mistakes in the DNN. This brings to our second contribution, as we can analyse communities and node similarities in the co-activation graph to understand which classes are similar from the DNN point of view. Also, this indicates that overlapping nodes between two classes might be recommended for tuning process in order to improve the model to better distinguish these classes. However, since our focus is to analyse the fidelity of the graph representation, evaluating the latter is beyond the scope of this work.

Fig. 4. Heatmaps from nodes correlated to *sneaker*, *trouser* and *pullover* respectively.

Fig. 5. On the left side is a heatmap from a node overlapping in classes *bag* and *trouser* and on right side a heatmap from node in *ankleboot*, *pullover* and *bag*.

4.4   Visualizing Neurons in Hidden Layers

So far we have only analysed the relationships between neurons in the hidden layer and neurons in the output layer. A third analysis was then conducted to see if the co-activation graph can help understand how nodes in the input layer impact the activation of neurons in the hidden layer. To do that, we used the edges between nodes in hidden layer and input layer to plot heatmaps for each neuron. Thus, colors in the heatmap are defined based on these edge weights.

Figure 4 shows heatmaps for three nodes strongly connected to classes *sneaker*, *trouser* and *pullover* respectively. We can see from these heatmaps that the inputs correlated to theses neurons are clearly related to their respective strongly connected classes. Figure 5 shows instead nodes that are overlapping between two or more classes. Note that for the nodes shared by only two classes (left) it is still possible to visualize some aspects of their classes (*bag* and *trouser*) in the heatmap. Yet, it is more difficult to identify which classes are related to node on right image in Fig. 5, since it is overlapping in three different classes. This is another evidence that overlapping nodes in the co-activation graph indicate neurons that are not being able to make a distinction between some classes in the DNN. Thus, they might be considered for a tuning process in order to improve the model to separate these classes and reduce possible classification errors.

This brings to our third contribution, as it is possible to visualize what parts of the image activate each neuron. Also, the heatmaps confirm the relationship between nodes and their most correlated classes.

To summarize the achieved results, we have first shown that by querying the co-activation graph we can detect which neurons have high impact over each class in the DNN. Then we found that by using community detection methods in this graph it is possible to detect groups of classes with similar semantics. In addition, we showed that classes with high similarity (shared nodes in the graph) are responsible for most mistakes in the model. Finally we used the relationships between neurons in hidden layers and input layer to visually analyse their roles.

5   Final Remarks

In this work we focused on the problem of explaining deep learning models. We proposed a method to extract knowledge from trained DNNs and to represent it as a graph, which is called co-activation graph. In the co-activation graph, nodes represent neurons in a DNN and weighted relationships indicate a statistical correlation between their activation values. Thus, it is a representation that connects neurons in any layer of the neural network, including hidden (convolutional and dense), input and the output layer through statistical correlations.

A preliminary evaluation was conducted to check whether the co-activation graph representation is compatible with the knowledge encoded in the DNN. Two experiments showed that by exploring the graph structure we can find neurons that have a high contribution for predicting some classes. This is crucial for better understanding: since these neurons highly impact the prediction values for some classes we can analyse them to explain decisions taken by the model.

Then, we showed that community detection methods over this graph can detect classes that are more similar from the DNN point of view, and that classes highly overlapping in the co-activation graph are responsible for most mistakes in the model. This is another interesting finding as it indicates that overlapping nodes in the graph can be considered for a tuning process for model improvement in order to better distinguish similar classes. At last we used the co-activation graph to show that relationships between hidden layers and the input layer can be used to visually explain what parts of the input layer activate each neuron in the hidden layer regardless of its type and location. By plotting heatmaps for each neuron we showed how this can help identifying their roles in the DNN.

Although we still consider these as initial results, they demonstrate the potential of using co-activation graphs to extract knowledge from DNNs. Our goal is to investigate this further and in a more formal way. As part of our next steps, we plan to consider other correlation coefficients to see if a non-linear correlation gives a better representation than Pearson coefficient. We also want to explore the notion of graph centrality to clarify whether nodes with a high centrality measure plays a specific role in the DNN. Community structure should also be better investigated especially in terms of their overlaps. Since some classes have shown to have a high similarity, methods for detecting overlapping communities might give better clusters than disjoint ones. Finally, experiments with bigger neural networks such as VGG16 CNN [15] should be conducted to assess if our approach can extract knowledge from more complex networks and how this knowledge can be used to improve the model.

Acknowledgements. The Insight Centre for Data Analytics is supported by Science Foundation Ireland under Grant Number 17/RC-PhD/3483.

References

1. Bartlett, E.B.: Self determination of input variable importance using neural networks. Neural Parallel Sci. Comput. **2**, 103–114 (1994)
2. Blondel, V., Guillaume, J.L., Lambiotte, R., Lefebvre, E.: Fast unfolding of communities in large networks. J. Stat. Mech. Theory Exp. (2008). https://doi.org/10.1088/1742-5468/2008/10/P10008
3. Chan, V., Chan, C.W.: Development and application of an algorithm for extracting multiple linear regression equations from artificial neural networks for nonlinear regression problems. In: 2016 IEEE 15th International Conference on Cognitive Informatics Cognitive Computing (ICCI*CC), pp. 479–488, August 2016
4. Erhan, D., Bengio, Y., Courville, A., Vincent, P.: Visualizing higher-layer features of a deep network. Technical report. Univeristé de Montréal, January 2009
5. Junque de Fortuny, E., Martens, D.: Active learning-based pedagogical rule extraction. IEEE Trans. Neural Netw. Learn. Syst. **26**, 2664–2677 (2015)
6. Garcia-Gasulla, D., et al.: Building graph representations of deep vector embeddings. CoRR abs/1707.07465 (2017). http://arxiv.org/abs/1707.07465

7. Garcia-Gasulla, D., et al.: An out-of-the-box full-network embedding for convolutional neural networks. In: 2018 IEEE International Conference on Big Knowledge (ICBK), pp. 168–175 (2018)
8. Kim, D.E., Lee, J.: Handling continuous-valued attributes in decision tree with neural network modeling. In: López de Mántaras, R., Plaza, E. (eds.) ECML 2000. LNCS (LNAI), vol. 1810, pp. 211–219. Springer, Heidelberg (2000). https://doi.org/10.1007/3-540-45164-1_22
9. Krishnan, R., Sivakumar, G., Bhattacharya, P.: A search technique for rule extraction from trained neural networks. Pattern Recogn. Lett. **20**(3), 273–280 (1999)
10. LeCun, Y., Cortes, C.: MNIST handwritten digit database (2010). http://yann.lecun.com/exdb/mnist/
11. Liu, B., Wei, Y., Zhang, Y., Yang, Q.: Deep neural networks for high dimension, low sample size data, pp. 2287–2293, August 2017. https://doi.org/10.24963/ijcai.2017/318
12. Mak, B., Blanning, R.W.: An empirical measure of element contribution in neural networks. IEEE Trans. Syst. Man Cyberne. Part C (Appl. Rev.) **28**(4), 561–564 (1998)
13. Mohamed, M.H.: Rules extraction from constructively trained neural networks based on genetic algorithms. Neurocomput. **74**(17), 3180–3192 (2011)
14. Simonyan, K., Vedaldi, A., Zisserman, A.: Deep inside convolutional networks: visualising image classification models and saliency maps. CoRR abs/1312.6034 (2013). http://arxiv.org/abs/1312.6034
15. Simonyan, K., Zisserman, A.: Very deep convolutional networks for large-scale image recognition. CoRR abs/1409.1556 (2014). http://arxiv.org/abs/1409.1556
16. Towell, G.G., Shavlik, J.W.: Extracting refined rules from knowledge-based neural networks. Mach. Learn. **13**(1), 71–101 (1993)
17. Xiao, H., Rasul, K., Vollgraf, R.: Fashion-MNIST: a novel image dataset for benchmarking machine learning algorithms (2017)

Question Formulation and Question Answering for Knowledge Graph Completion

Maria Khvalchik$^{(\boxtimes)}$, Christian Blaschke, and Artem Revenko

Semantic Web Company, Vienna, Austria
{maria.khvalchik,christian.blaschke,artem.revenko}@semantic-web.com

Abstract. Knowledge graphs contain only a subset of what is true. Following recent success of Question Answering systems in outperforming humans, we employ the developed tools to complete knowledge graph. To create the questions automatically, we explore domain-specific lexicalization patterns. We outline the overall procedure and discuss preliminary results.

Keywords: Knowledge graph completion · Question formulation · Question Answering · Link prediction

1 Introduction

Knowledge graphs (KGs) contain knowledge about the world and provide a structured representation of this knowledge. Current KGs contain only a small subset of what is true in the world [7]. There are different types of information that could be incomplete, for example, incomplete set of entities, incomplete set of predicates, or missing links between existing entities. Different types of incompleteness are usually addressed with different methods, for example, Named Entity Recognition is successfully applied to find new entities of given classes [9]. In this work we consider the latter problem of link prediction, i.e. finding triples (subject s, predicate p, objects o), where p is defined in the schema of the KG and s and o are known instance contained in the existing KG.

The approaches for knowledge extraction can roughly be subdivided into two classes:

Leveraging the knowledge from the existing KG

- The rule induction methods [3,6] learn rules over KG that capture patterns in data. In a generic domain one can learn, for example, that a

---

This work has been partially funded by the project LYNX. The project LYNX has received funding from the European Union's Horizon 2020 research and innovation programme under grant agreement no. 780602. More information is available online at http://www.lynx-project.eu.

G. Anderst-Kotsis et al. (Eds.): DEXA 2019 Workshops, CCIS 1062, pp. 166–171, 2019.
https://doi.org/10.1007/978-3-030-27684-3_21

person has a home address or that a consumer good has a price. These rules help to identify potential gaps in an incomplete KG. In order to fill in the gaps one needs to verify the veracity of the potential new triple.

- Embeddings project symbolic entities and relations into continuous vector space. The vector arithmetic is used to predict new links [8, 11].

Extracting the knowledge from other sources

- Transformation of the information from some structured source[1].
- Relation Extraction methods [10] employ trained models to recognize triples in the text and add those triples to the existing KG.

In this work we consider combining both approaches. Given a KG we use pattern mining and heuristics to identify potential gaps in the KG. Then we employ a Question Answering (QA) framework to find and/or verify new triples.

2 Approach

After applying graph pattern mining techniques we obtain generalized graph patterns [1]. These patterns help identifying missing links. We aim at restoring missing links and for this purpose formulate a query that is in the simplest case equal to a pair (s, p). The task is to find objects O such that $\{(s, p, o) \mid o \in O\}$ is a set of valid triples. In case of having triples, the task is to verify the provided triples. Our approach consists of the following steps:

1. Question formulation,
2. Retrieving documents potentially containing answers from the corpus,
3. Employing QA over documents to get candidate answers and their scores,
4. Choosing correct answers.

Question Formulation. The goal is to go from a pair (s, p) to such a question that a correct answer o defines a valid triple (s, p, o).

In order to articulate this natural language question q we employ lexicalization techniques [5]. Specifically, we intend to learn general question templates following with the lexicalization of domain-specific predicates. The overall schema is presented in Fig. 1 and described below.

---

[1] https://www.w3.org/TR/r2rml/.

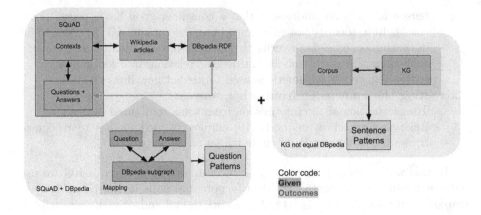

Fig. 1. Question and sentence patterns formulation (Color figure online)

The Question Patterns (QPs) are learned from SQuAD[2] and DBpedia[3] and are generalized ways to construct a question parameterized by predicate p, subject s, and other bound variables in a query, for example:

$$QP_1(s, p) = <Wh\text{-}word(p) \ was \ Lex(s) \ Lex(p)>,$$

where $Wh\text{-}word(x)$ is a function of predicate that can output values from $\{$ *What, When, Where,* ...$\}$, *Lex* is a function from property or instance to natural language string, for example: $Lex(\texttt{dbo:birthPlace}) = $ "born". Examples of the natural language questions generated by QP_1 are:

- $QP_1(\texttt{dbr:Homer}, \texttt{dbo:birthPlace}) = $ "Where was Homer born?"
- $QP_1(\texttt{dbr:Homer}, \texttt{dbo:birthDate}) = $ "When was Homer born?"

For Sentence Patterns (SPs) the input is a set of pairs $\{(t, sg_t) \mid t \in C\}$, where t is a text from a corpus C, sg_t is a subgraph of DBpedia aligned to t, i.e. t and sg_t contain the same knowledge expressed in natural language and RDF, correspondingly. The SPs are used to generate the *Lex* and *Wh-word* functions.

In the KG we might assume that classes are mapped to DBpedia classes, and properties are defined with domain and range. Hence, we can reuse *Wh-word* function as it is generated from DBpedia dataset and takes as input the predicate's range from DBpedia.

The output of the whole procedure is a question in natural language.

Retrieving Documents. Given a natural language question q we use query expansion techniques [2] and formulate a search query to retrieve relevant documents from the corpus. For this task we use an indexer which job is to collect, parse and store data to facilitate fast and accurate information retrieval.

---

[2] https://rajpurkar.github.io/SQuAD-explorer/.
[3] https://wiki.dbpedia.org/.

Question Answering. Our end-to-end QA system accepts a natural language question and a set of documents as input and outputs a set of pairs (answer, confidence score). The score indicates the confidence of the system that the answer is correct. The QA systems are usually computationally demanding, therefore it is not feasible to send the whole corpus as an input to the QA system.

We use an implementation of BERT [4] that we trained on a large question-answer pairs dataset SQuAD. BERT is a deep learning system that outperforms previous approaches and reaches close to human performance in QA tasks.

Verification and Integration. Given the answers produced by the QA system we try to match each answer to an instances in the KG. If an answer could not be matched then it is discarded. For the undiscarded answers we check if the class of the matched instances complies with the range restrictions of the predicate definition in the schema. If the range restriction is satisfied we add the triple (subject, predicate, matched instances) to the KG.

In the special case when we know that at most one triple is allowed for a pair (subject, predicate) we choose the answer yielding the highest score of the QA system if it satisfies the range restriction.

3 Preliminary Results and Conclusion

The first used corpus is a collection of over 3,300 biomedical paper abstracts from "PubMed"[4] containing manually queried term "BRCA1". Questions and first 3 answers are provided in the Table 1. For the first question: "Which diseases does niraparib treat?" it is possible to check the following triple (niraparib, treatsDisease, BRCA-mutant ovarian cancer). If this triple fails the verification stage, we continue to verify other candidates for their existence. In this case the answer satisfies the verification as "BRCA-mutant ovarian cancer" is indeed a disease.

In all three examples we asked what kind of disease is treated by such drugs as niraparib, rucaparib, and olaparib. All these drugs are anti-cancer agents, the system can successfully match these drugs with treating cancer. Moreover, the system is able to identify the specific types of cancer and additional details.

In the case of Rucaparib the approved indication for this drug is "Advanced Ovarian Cancer"[5]. The answer retrieved by the system is "advanced ovarian and breast cancer". "Advanced ovarian cancer" is indeed correct, and interestingly, by looking at the document where the answer came from, it follows: "We investigated the efficacy and safety of single-agent rucaparib in germline (g) BRCA mutation carriers with advanced breast and ovarian cancers"[6]. This shows that the system also retrieves other facts that are correct but not within the narrow range of the KG we used as a reference. DrugBank only contains FDA[7] approved

---

[4] https://www.ncbi.nlm.nih.gov/pubmed/.

[5] https://www.drugbank.ca/drugs/DB12332.

[6] https://www.ncbi.nlm.nih.gov/pubmed/27002934.

[7] https://www.fda.gov.

Table 1. Questions and answers.

Question	Answers, confidence score
PubMed dataset with term "BRCA1"	
Which diseases does niraparib treat?	1. BRCA-mutant ovarian cancer, 0.259
	2. Tumors with Defective Homologous Recombination, 0.226
	3. Ovarian cancer, 0.223
Which diseases does rucaparib treat?	1. Advanced ovarian and breast cancer, 0.204
	2. Nausea, vomiting, asthenia/fatigue, anemia and transient transaminitis, 0.136
	3. Patients with Deleterious BRCA Mutation-Associated Advanced Ovarian Cancer, 0.114
Which diseases does olaparib treat?	1. Ovarian cancer and BRCA mutations, 0.446
	2. Ovarian cancer, 0.111
	3. Specific forms of ovarian cancer and BRCA mutations, 0.063
PubMed dataset with term "rs1045642"	
What is rs1045642 associated with?	1. Risk of mucositis, 0.366
	2. Increased abdominal fat mass and decreased appendicular skeletal muscle mass, 0.236
	3. Time to allograft failure, 0.211
What does rs1045642 modulate?	1. Office Blood Pressure, 0.780
	2. Blood Pressure, 0.081
	3. Office Blood Pressure at 1-Year Post Kidney Transplantation, 0.061

indications whereas the document the answer came from refers to studies that are currently performed on additional indications. Though the relevancy of the facts is use case specific, this example shows the limitations of KGs and their completeness for one scenario but not another.

The second dataset is a collection of paper abstracts containing manually queried term "rs1045642", the identifier of a single-nucleotide polymorphism (SNP) in the human genome. We intend to observe data about mutations, the KG of which we expect not to be complete as the database of mutations is not up-to-date. To add such frequently updated data one should inspect corresponding literature. That being said, it is clear that an automated system could come as a benefit.

4   Conclusion

We considered an important and practically relevant task of link prediction in KGs. In our approach we combine link prediction techniques with QA system to extract concealed knowledge from a text corpus and to formulate new triples. The first experiments show promising results for domain-specific datasets.

References

1. Abedjan, Z., Naumann, F.: Improving RDF data through association rule mining. Datenbank-Spektrum **13**(2), 111–120 (2013)
2. Bhogal, J., Macfarlane, A., Smith, P.: A review of ontology based query expansion. Inf. Process. Manag. **43**(4), 866–886 (2007)
3. d'Amato, C., Staab, S., Tettamanzi, A.G.B., Minh, T.D., Gandon, F.: Ontology enrichment by discovering multi-relational association rules from ontological knowledge bases. In: ACM 31, pp. 333–338 (2016)
4. Devlin, J., Chang, M., Lee, K., Toutanova, K.: BERT: pre-training of deep bidirectional transformers for language understanding. CoRR abs/1810.04805 (2018)
5. Ell, B., Harth, A.: A language-independent method for the extraction of RDF verbalization templates. In: INLG 2014, pp. 26–34 (2014)
6. Ho, V.T., Stepanova, D., Gad-Elrab, M.H., Kharlamov, E., Weikum, G.: Rule learning from knowledge graphs guided by embedding models. In: Vrandečić, D., et al. (eds.) ISWC 2018. LNCS, vol. 11136, pp. 72–90. Springer, Cham (2018). https://doi.org/10.1007/978-3-030-00671-6_5
7. Ji, G., He, S., Xu, L., Liu, K., Zhao, J.: Knowledge graph embedding via dynamic mapping matrix. In: ACL 2015, Long Papers,, vol. 1, pp. 687–696 (2015)
8. Lin, Y., Liu, Z., Sun, M., Liu, Y., Zhu, X.: Learning entity and relation embeddings for knowledge graph completion. In: AAAI 29, pp. 2181–2187 (2015)
9. Sanchez-Cisneros, D., Aparicio Gali, F.: UEM-UC3M: an ontology-based named entity recognition system for biomedical texts. In: SemEval, pp. 622–627. Association for Computational Linguistics (2013)
10. Schutz, A., Buitelaar, P.: *RelExt*: a tool for relation extraction from text in ontology extension. In: Gil, Y., Motta, E., Benjamins, V.R., Musen, M.A. (eds.) ISWC 2005. LNCS, vol. 3729, pp. 593–606. Springer, Heidelberg (2005). https://doi.org/10.1007/11574620_43
11. Wang, Z., Zhang, J., Feng, J., Chen, Z.: Knowledge graph embedding by translating on hyperplanes. In: AAAI 2014, pp. 1112–1119 (2014)

The Use of Class Assertions and Hypernyms to Induce and Disambiguate Word Senses

Artem Revenko[✉] and Victor Mireles

Semantic Web Company, Vienna, Austria
{artem.revenko,victor.mireles-chavez}@semantic-web.com

Abstract. With the spread of semantic technologies more and more companies manage their own knowledge graphs (KG), applying them, among other tasks, to text analysis. However, the proprietary KGs are by design domain specific and do not include all the different possible meanings of the words used in a corpus. In order to enable the usage of these KGs for automatic text annotations, we introduce a robust method for discriminating word senses using sense indicators found in the KG: types, synonyms and/or hypernyms. The method uses collocations to induce word senses and to discriminate the sense included in the KG from the other senses, without the need for information about the latter, or the need for manual effort. On the two datasets created specially for this task the method outperforms the baseline and shows accuracy above 80%.

Keywords: Thesaurus · Controlled vocabulary ·
Word Sense Induction · Entity linking · Named entity disambiguation

1 Introduction

Since about 80% of the words used in everyday English utterances are ambiguous (Rodd et al. 2002), even words included in domain-specific vocabularies can appear in several senses in a given corpus. Therefore, for information retrieval to successfully provide services like search or recommendation, it is necessary to distinguish among the different senses of words, both in queries and in corpora, regardless of domain-specificity. This task is known as word sense disambiguation (WSD). If the target words are domain-specific entities, the task is called named entity disambiguation (NED), which is the focus of this work.

In domain-specific applications it is possible for experts to define the domain vocabulary in a controlled manner. In this work, we define a *controlled vocabulary* as a finite and well specified set of *entities* from a specific domain. Each entity has, among other attributes, a set of one or more *labels*, each of which may be a word or words habitually used in a (natural) language, an acronym or even

© Springer Nature Switzerland AG 2019
G. Anderst-Kotsis et al. (Eds.): DEXA 2019 Workshops, CCIS 1062, pp. 172–181, 2019.
https://doi.org/10.1007/978-3-030-27684-3_22

an arbitrary sequence of symbols. Having several labels for a single entity, a controlled vocabulary groups different synonyms or alternative spellings, increasing the focus of text analysis methods on meaning.

Controlled vocabularies can be enriched with semantic information, resulting in a KG (Ehrlinger and Wöß 2016): a tuple consisting of i. a controlled vocabulary of entities, ii. a schema consisting of classes and relations, iii. assertions about entities belonging to classes, entities participating in relations with other entities, and assignments of attributes to entities One example of an attribute that an entity posses is a label. A particular example of relations are the mutually inverse relations of *hyponym* and *hypernym*. The hypernym of an entity x is defined as an entity y whose meaning includes the meaning of x. A KG containing entities, their sets of labels (synonym sets) and hierarchical hyponym-hypernym relations between the entities constitutes a *thesaurus*.

Example 1. The Medical Subject Headings (MeSH)[1] thesaurus is a domain-specific controlled vocabulary produced by the National Library of Medicine and used for indexing, cataloging, and searching for biomedical and health-related information and documents. The associated service MeSHonDemand [2] identifies MeSH terms in the submitted text.

The MeSH thesaurus includes an entity *Vertigo* (https://id.nlm.nih.gov/mesh/D014717.html) which is hyponym of entities with the labels *Neurologic Manifestations* and *Vestibular Diseases*. Yet, in the phrase taken from the abstract of (Berman 1997):

> The author presents an interpretation of Hitchcock's 'Vertigo', focusing on the way in which its protagonist's drama resonates with the analyst's struggle with deep unconscious identifications...

the word *Vertigo* does not refer to the MeSH entity.

1.1 Problem Statement

We present in this work a method that tackles the domain-specific case of NED by using a KG, without requiring all possible senses of a word to be contained in it. This makes it specially useful in the industrial environment, where usually only small thesauri are available. Specifically, we take as an input a corpus, a KG, and an entity from the KG, one of whose labels is found throughout the corpus possibly in different senses. We call this label the *target label*. The problem is to distinguish, for each document in the corpus, whether the target label is used in the in-domain sense, i.e., in the sense corresponding to the entity in the KG, or not. Without loss of generality we consider only the case when the target label has only one in-domain sense. Thus, the end result is a partition of the corpus into two disjoint collections: "this" and "other". The collection "this" contains the documents that feature the target label in the in-domain sense.

---

[1] https://meshb.nlm.nih.gov/ (visited on 05.03.2019).

[2] https://meshb.nlm.nih.gov/MeSHonDemand (visited on 05.03.2019).

In the introduced method we make use of various types of information depending on what is provided in controlled vocabulary: types of entities, synonyms, hypernyms, etc. We call them jointly *sense indicators*.

The contributions of this work are the following:

- We introduce a method for Word Sense Induction (WSI) with the usage of sense indicators.
- We introduce a pipelined workflow to discriminate between in-domain and not-in-domain senses of the target entity, by utilizing sense indicators.
- We prepare and carry out an experiment that resembles a real world use case.

2   Related Work

The problem of word sense disambiguation has attracted much attention for several decades now. We refer the reader to (Navigli 2012) for a thorough review. Among them, several methods have been developed to exploit knowledge encoded in KGs. These are, in general, based on static knowledge graphs (KG) that are computed a priori (Rao et al. 2013; Ratinov et al. 2011), usually with great effort, such as WordNet (Navigli et al. 2011; Zhong and Ng 2010), and usually assume that the graph includes all senses of the target word, and of most other words that appear in the corpus. To disambiguate a word in a given context, they traverse the KG starting from entities matching the word and its context (e.g., with a random walk (Agirre et al. 2014)) and the sense that is returned is that which is *closer* in the graph to the starting points.

Many WSD methods, whether incorporating external knowledge or not, include Word Sense Induction (WSI) stage. This stage consists of finding (inducing) the senses present in unannotated corpus. In the terms described above, the input to this task is a corpus and a target label. The outcome is an enumeration of all the senses found for the target label. The WSI problem has been approached from many angles. The first is the so-called sense embedding approach (Li and Jurafsky 2015; Neelakantan et al. 2015). Methods of this sort embed each label into a multi-dimensional space. Then for each context an embedding can be computed from the embeddings of the labels in the context. The produced context embeddings are then clustered to determine the different senses of each word. We should mention that, in our experience, these methods fail in cases where two senses of a word lead to very similar contexts, such as "Americano" which can refer to either a cocktail or a coffee.

The second common approach to the WSI task which is of interest here, is to analyze the text and extract from it collocation graphs: graphs whose nodes are words found in the text close to the target word, and whose edges are weighted to reflect the strength of this collocation. This has the advantage that previously unknown senses can be induced and described with the help of collocations. To weight the edges of the graph, conditional probabilities (Véronis 2004), Dice scores (Di Marco and Navigli 2013) or word co-occurrences relative to a reference corpus (Klapaftis and Manandhar 2008) have been used, as well as discrete features derived from the syntactic use of each word (e.g., (Panchenko et al.

2017)). Within these weighted graphs, graph algorithms identify a collection of sets of nodes, each of which corresponds to a sense. Once these sets are identified, WSI can be carried out by clustering the contexts where the target word appears, for example, via graph clustering algorithms (e.g., in (Di Marco and Navigli 2013; Dorow and Widdows 2003; Panchenko et al. 2017)). Another approach, built specifically for inducing senses, is HyperLex (Véronis 2004), which defines senses as hubs (highly connected nodes) in the collocation graph along with their immediate neighbors. To identify senses, HyperLex sorts the nodes of the collocation graph by degrees. Senses are induced by taking and removing one by one the hubs from list, along with their immediate neighbors. In this paper we use a variant of HyperLex introduced in (Di Marco and Navigli 2013) with PageRank scores in place of degrees.

We note that in the case of collocation graphs, there is no explicit description of what a particular sense is. This lack of interpretability of a sense is overcome in this work by relating at least one of the collocation-derived senses with an entity in the KG.

3   Method

In this section we introduce a method to solve the task introduced in Sect. 1.1. We rely on the "one sense per document" assumption (Yarowsky 1995) and on the assumption that the sense of the entity can be unambiguously deduced from its sense indicators.

Our method consists of 2 steps:

1. WSI, illustrated in Fig. 1; the outcome is a set of senses with one distinguished in-domain sense.
2. WSD, i.e., classification of each occurrence of the target word into one of the senses.

3.1   WSI with Hypernyms

For the WSI task we use HyperLex (Véronis 2004), implementing it in a similar way to (Di Marco and Navigli 2013). However, we also introduce sense indicators in the WSI process, therefore we will denote the new modification as *HyperHyperLex*[3]. This method takes sense indicators into account already at the stage of graph clustering. This guarantees that – the thesaural meaning of the target word is captured in a single sense and – the level of granularity of the sense is defined by the structure of the thesaurus. The process is presented in Fig. 1. HyperHyperLex can be divided into four steps: **first** we compute the graph of collocations between all the words in the text. In Fig. 1 this phase is represented as a graph of collocations, where "*L*" stands for target label. In the **second** step we compute the PageRank of the nodes in the graph. In Fig. 1 we represents the

---

[3] The additional "hyper" stems from "hypernym" as one of the most useful sense indicators.

nodes having larger values of PageRank with thicker lines. In the **third** step we introduce the sense indicators into the process and for each node we compute the following measure

$$m(n) := PR(n) * \left(CO(L, n) + \frac{\sum_{s \in S} CO(s, n)}{|S|} \right),$$

where PR stands for PageRank, S is the set of sense indicators, and CO is the collocation measures. We use a variant of Dice Scores (Dice 1945) as a word-association measure to grab collocations.

In the **fourth** step, nodes are assigned to different clusters. The node n with the largest $m(n)$ is taken as a *hub*. Next we build a cluster around the hub by assessing the *involvement* of every other node with the hub, as the sum of the Dice scores between the node and the hub and its direct neighbors. The new hub and its neighbors are cut out from the graph and the procedure is repeated until there are no more nodes with large $m(n)$ values. When a not-in-domain sense is induced the set of sense indicators S is empty and $m(n) = PR(n) * CO(L, n)$.

After this process, each node has a weight denoting its membership in a cluster. This weight is used in the disambiguation step for classifying the occurrences of the target entity. The weight is proportional to

$$w(n) \sim PR(n) * I(n),$$

where I stands for involvement of the node in said cluster.

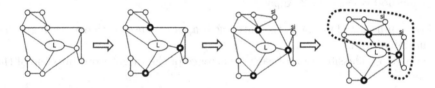

Fig. 1. HyperHyperLex. L stands for the target label, si stand for sense indicator. Thick nodes identify the hubs, dashed line marks the hypernym-induced sense.

In the preliminary tests using original HyperLex (without sense indicators) we found that several senses corresponding to the in-domain sense could be induced. However, none of the senses captured the in-domain sense completely, resulting in misclassification. With the help of the sense indicators it has become possible to capture the in-domain sense in a single sense and take into account the decisions of the data architect reflected in the KG.

3.2 NED

The result of the previous WSI step is a set of key-value dictionaries, where

- each dictionary corresponds to an induced sense,

- the keys are the words that belong to the sense cluster,
- the values are the weights $w(n)$ of the words with respect to the sense.

In order to disambiguate an occurrence of the target label we first extract its context. A context is a set of words surrounding the target label, for example, 10 words before and 10 words after the target. Then for each sense we take the corresponding dictionary and sum up the weights of all the words that are found in the context and in the sense cluster. Finally, we choose the sense with the highest aggregate score.

4   Evaluation

4.1   Cocktails and MeSH

We handcraft 2 datasets[4] specifically for the task of WSID with one known sense and several unknown ones: cocktails and MeSH (Nelson 2009).

Data. In the cocktails example the entities are taken from the "All about cocktails"[5] thesaurus. The thesaurus contains various cocktail instances and ingredients. We have only used ambiguous cocktail names for this experiment. We use the entity scheme name "cocktail" and the broaders of the target entity as sense indicators.

In the MeSH example we use the MeSH[6] thesaurus. For each entity, we take as sense indicators the labels of its broader entities, specifically those marked as preferred.

As we only use unigrams in our code, we split the compound labels into unigrams.

We used the Wikilinks dataset (Singh et al. 2012) to extract the corpora. The dataset contains documents and the links to the Wikipedia pages inside the documents. The texts contain mistakes, which makes them particularly suitable for simulating a real world use case. The corpora contain duplicates. The data is used as is, without any cleaning.

First we identify those labels in a thesaurus that are ambiguous and then we collected the texts that mention those labels in different sense. The preprocessing phase includes removing stopwords, removing rare and frequent words.

We collect 13 corpora with a total of 1227 texts for cocktails, see Table 1 for more details. We collect 8 corpora with a total of 784 texts for MeSH labels, see Table 2 for more details.

Results. In the experiment we classify the word occurrences into two categories "this" and "other". We should note that many considered words have more than 2 meanings, all the non-thesaural meanings fall into the category "other". For

---

[4] github.com/artreven/thesaural_wsi (visited on 05.03.2019).

[5] vocabulary.semantic-web.at/cocktails (visited on 05.03.2019).

[6] www.nlm.nih.gov/mesh/ (visited on 05.03.2019).

the baseline everything is classified into the most popular category. This baseline
is known to be challenging. In practice there is no guarantee that the in-domain
sense would be the most popular, therefore this baseline is better than the results
one could expect in practice without WSID.

The results for cocktails are presented in Table 1. Observations:

- As can be seen from results even the high number of the real senses does not
  prevent the method from showing high accuracy.
- With large corpora the accuracy is high due to better WSI.
- Even if the number of in-domain mentions is very low ("Cosmopolitan", "B-
  52") the accuracy remains high. We assume that the well represented "other"
  senses can be induced accurately and the use of hypernyms improves the
  induction of the in-domain sense.
- The worst results are obtained for the corpora where the in-domain sense
  (cocktail) is the dominant: Vesper, Margarita, Martini. Since the other
  sense(s) are underrepresented there is a risk of getting several senses capturing the individual context. In such senses many general-purpose (not sense-
  specific) collocations may be present, as a result these senses may score higher
  in general contexts.

The results for MeSH are presented in Table 2. These results are worse than of the
"cocktails" dataset. We observe due to the methodology of corpus acquisition,
namely blog texts, there are very little texts mentioning entities - symptoms and
diseases - in the domain-specific MeSH sense.

The averaged results are presented the last 2 lines of respective tables. In
both cases HyperHyperLex outperforms the challenging baseline. The difference
is even larger when the individual texts are taken into account (micro average),
because the method benefits from having a larger corpus.

We performed WSI using HyperLex for comparison. The results for some
words are comparable or even better, however for other words the accuracy
drops significantly.

4.2   SemEval 2013 Task 13

We benchmark our method on the data provided by SemEval 2013 challenge in
Task 13 [7]. The task seeks to identify the different senses (or uses) of a target
label in a given text in an automatic and fully-unsupervised manner. Target
words are not Named Entities, hence the use case is slightly different and more
challenging than the original motivating use case. Both our method and the task
are "motivated by the limitations of traditional knowledge-based & supervised
Word Sense Disambiguation methods, such as their limited adaptation to new
domains and their inability to detect new senses (uses) not present in a given
dictionary".

In the induction step we used the UkWaC corpus[8] without PoS tags, therefore
for words like "book" the induced senses were a mixture of verbal and noun

---

[7] www.cs.york.ac.uk/semeval-2013/task13/ (visited on 05.03.2019).

[8] wacky.sslmit.unibo.it/doku.php (visited on 05.03.2019).

Table 1. Accuracy of discriminating cocktail names using hypernyms

Cocktail name	Number of senses	Number of texts	Cocktail sense texts	Baseline accuracy	HyperLex accuracy	HyperHyperLex accuracy
Americano	2	45	13	0.711	0.844	0.889
Aviation	2	27	17	0.63	0.37	0.852
B-52	2	111	8	0.928	0.946	0.982
Bellini	3	95	42	0.558	0.737	0.968
Bloody Mary	6	352	109	0.69	0.827	0.935
Cosmopolitan	3	125	10	0.92	0.928	0.952
Grasshopper	4	39	13	0.667	0.744	0.974
Manhattan	4	262	70	0.733	0.851	0.885
Margarita	2	41	36	0.878	0.878	0.561
Martini	2	38	27	0.711	0.711	0.711
Mimosa	3	45	18	0.6	0.4	0.733
Tequila Sunrise	2	16	10	0.625	0.938	0.812
Vesper	2	31	24	0.774	0.839	0.677
Macro average		1227	397	0.725	0.784	0.841
Micro average				0.737	0.821	0.896

Table 2. Accuracy of discriminating MeSH labels using hypernyms

MeSH label	Number of senses	Number of texts	MeSH sense texts	Baseline accuracy	HyperLex	HyperHyperLex
Amnesia	2	31	18	0.581	0.806	0.806
Dengue	2	84	59	0.702	0.298	0.583
Warts	2	14	9	0.643	0.643	0.643
Delirium	2	135	98	0.726	0.711	0.556
Iris	5	150	29	0.807	0.433	0.720
Kuru	2	69	35	0.507	0.696	0.957
Amygdala	2	64	42	0.656	0.578	0.656
Vertigo	5	237	43	0.819	0.768	0.865
Macro average		784	333	0.680	0.617	0.723
Micro average				0.735	0.621	0.739

senses. Obviously, the results could be improved by taking the PoS tags into account, no modification of the method is required. We aimed at remaining language-agnostic and did not want to employ language-specific PoS taggers.

For each word the hypernyms and synonyms from WordNet senses were taken as the sense indicators. The top 3 results were kept for each context. The results of evaluation of our method are presented in Table 3, the results of the systems participating in the challenge are given at https://www.cs.york.ac.uk/semeval-2013/task13/index.php%3Fid=results.html (Visited on 05.03.2019).

Table 3. Results of SemEval 2013 Task 13 challenge

	Jaccard Index	Positionally-Weighted Tau	Weighted NDCG Recall	Fuzzy NMI	Fuzzy B-Cubed Fscore
All	0.193	0.603	0.369	0.092	0.393
All multi-senses	0.386	0.632	0.319	0.069	0.079

5   Conclusion

We have introduced a method to automatically discriminate between in-domain and non-in-domain usage of words. The method requires neither the knowledge of all sense of the target word, nor the number of senses and does not make use of external resources except for controlled vocabulary, which is considered an input parameter.

In the two experiments on created datasets the new method has outperformed the baseline and has shown an accuracy of about 0.9 and 0.75, respectively. Though the method is designed to disambiguate named entities, it performs comparable to other more specialized systems in disambiguation of general contexts provided at SemEval2013 Task 13.

The method could be applicable in all types of entity matching tools and improve their matching results. An example of such tool is meshb.nlm.nih. gov/MeSHonDemand.

Acknowledgements. This work was supported in part by the H2020 project Prêt-á-LLOD under Grant Agreement number 825182.

References

Agirre, E., de Lacalle, O.L., Soroa, A.: Random walks for knowledge-based word sense disambiguation. Comput. Linguist. **40**(1), 57–84 (2014)

Berman, E.: Hitchcock's vertigo: the collapse of a rescue fantasy. Int. J. Psycho-Anal. **78**, 975–988 (1997)

Di Marco, A., Navigli, R.: Clustering and diversifying web search results with graph-based word sense induction. Comput. Linguist. **39**(3), 709–754 (2013)

Dice, L.R.: Measures of the amount of ecologic association between species. Ecology **26**(3), 297–302 (1945)

Dorow, B., Widdows, D.: Discovering corpus-specific word senses. In: Proceedings of the Tenth Conference on European Chapter of the Association for Computational Linguistics, vol. 2, pp. 79–82. Association for Computational Linguistics (2003)

Ehrlinger, L., Wöß, W.: Towards a definition of knowledge graphs. SEMANTiCS (Posters, Demos, SuCCESS), 48 (2016)

Klapaftis, I.P., Manandhar, S.: Word sense induction using graphs of collocations. In: ECAI, pp. 298–302 (2008)

Li, J., Jurafsky, D.: Do multi-sense embeddings improve natural language understanding? arXiv preprint arXiv:1506.01070 (2015)

Navigli, R.: A quick tour of word sense disambiguation, induction and related approaches. In: Bieliková, M., Friedrich, G., Gottlob, G., Katzenbeisser, S., Turán, G. (eds.) SOFSEM 2012. LNCS, vol. 7147, pp. 115–129. Springer, Heidelberg (2012). https://doi.org/10.1007/978-3-642-27660-6_10

Navigli, R., Faralli, S., Soroa, A., de Lacalle, O., Agirre, E.: Two birds with one stone: learning semantic models for text categorization and word sense disambiguation. In: Proceedings of the 20th ACM International Conference on Information and Knowledge Management, pp. 2317–2320. ACM (2011)

Neelakantan, A., Shankar, J., Passos, A., McCallum, A.: Efficient non-parametric estimation of multiple embeddings per word in vector space. arXiv preprint arXiv:1504.06654 (2015)

Nelson, S.J.: Medical terminologies that work: the example of mesh. In: 2009 10th International Symposium on Pervasive Systems, Algorithms, and Networks, pp. 380–384 (2009)

Panchenko, A., Ruppert, E., Faralli, S., Ponzetto, S.P., Biemann, C.: Unsupervised does not mean uninterpretable: the case for word sense induction and disambiguation. Association for Computational Linguistics (2017)

Rao, D., McNamee, P., Dredze, M.: Entity linking: finding extracted entities in a knowledge base. In: Poibeau, T., Saggion, H., Piskorski, J., Yangarber, R. (eds.) Multi-source, Multilingual Information Extraction and Summarization. Theory and Applications of Natural Language Processing, pp. 93–115. Springer, Heidelberg (2013). https://doi.org/10.1007/978-3-642-28569-1_5

Ratinov, L., Roth, D., Downey, D., Anderson, M.: Local and global algorithms for disambiguation to wikipedia. In: Proceedings of the 49th Annual Meeting of the Association for Computational Linguistics: Human Language Technologies, vol. 1, pp. 1375–1384. Association for Computational Linguistics (2011)

Rodd, J., Gaskell, G., Marslen-Wilson, W.: Making sense of semantic ambiguity: semantic competition in lexical access. J. Mem. Lang. **46**(2), 245–266 (2002)

Singh, S., Subramanya, A., Pereira, F., McCallum, A.: Wikilinks: a large-scale cross-document coreference corpus labeled via links to Wikipedia. Technical report UM-CS-2012-015, University of Massachusetts, Amherst (2012)

Véronis, J.: Hyperlex: lexical cartography for information retrieval. Comput. Speech Lang. **18**(3), 223–252 (2004)

Yarowsky, D.: Unsupervised word sense disambiguation rivaling supervised methods. In: Proceedings of the 33rd Annual Meeting on Association for Computational Linguistics, pp. 189–196. Association for Computational Linguistics (1995)

Zhong, Z., Ng, H.T.: It makes sense: a wide-coverage word sense disambiguation system for free text. In: Proceedings of the ACL 2010 System Demonstrations, pp. 78–83. Association for Computational Linguistics (2010)

On Conditioning GANs to Hierarchical Ontologies

Hamid Eghbal-zadeh[1] , Lukas Fischer[2](✉) , and Thomas Hoch[2]

[1] LIT AI Lab & Institute of Computational Perception,
Johannes Kepler University Linz, Altenberger Straße 69, Linz, Austria
hamid.eghbal-zadeh@jku.at
[2] Software Competence Center Hagenberg GmbH (SCCH),
Softwarepark 21, 4232 Hagenberg, Austria
{lukas.fischer,thomas.hoch}@scch.at

Abstract. The recent success of Generative Adversarial Networks (GAN) is a result of their ability to generate high quality images given samples from a latent space. One of the applications of GANs is to generate images from a text description, where the text is first encoded and further used for the conditioning in the generative model. In addition to text, conditional generative models often use label information for conditioning. Hence, the structure of the meta-data and the ontology of the labels is important for such models. In this paper, we propose Ontology Generative Adversarial Networks (O-GANs) to handle the complexities of the data with label ontology. We evaluate our model on a dataset of fashion images with hierarchical label structure. Our results suggest that the incorporation of the ontology, leads to better image quality as measured by Fréchet Inception Distance and Inception Score. Additionally, we show that the O-GAN better matches the generated images to their conditioning text, compared to models that do not incorporate the label ontology.

Keywords: Generative Adversarial Networks ·
Text-to-image synthesis · Ontology-driven deep learning

1 Introduction

Text-to-image synthesis is a challenging task where the details about the respective images are provided in the text format. The details of the generated image should best fit to the explanation provided in the text description, while maintaining a high-level of image detail fidelity. Generative adversarial networks (GANs) have proven to be a very powerful method to tackle this task [2,5]. In

This work was supported by Austrian Ministry for Transport, Innovation and Technology, the Ministry of Science, Research and Economy, and the Province of Upper Austria in the frame of the COMET center SCCH.
H. Eghbal-zadeh, L. Fischer, T. Hoch—Equal contribution.

© Springer Nature Switzerland AG 2019
G. Anderst-Kotsis et al. (Eds.): DEXA 2019 Workshops, CCIS 1062, pp. 182–186, 2019.
https://doi.org/10.1007/978-3-030-27684-3_23

[6] it has been shown that hierarchical model training by means of an ontology helps to learn more discriminant high-level features for fashion image representations. We adopt this strategy for generative models and show in this paper how a two-tier fashion category taxonomy can be leveraged to improve GANs training for fashion image generation from text. The recently organized Fashion-Gen challenge [8][1] provides a perfect test bed for the evaluation of novel methods for text-to-image synthesis. The provided dataset consists of 293.008 images, with 48 main and 132 fine-grained categories as well as a detailed description text. An extract of this ontology is visualized in Fig. 1. To handle the complexities of the Fashion-Gen challenge, we propose Ontology Generative Adversarial Networks (O-GANs) for high-resolution text-conditional fashion image synthesis. We detail the O-GAN in the following section.

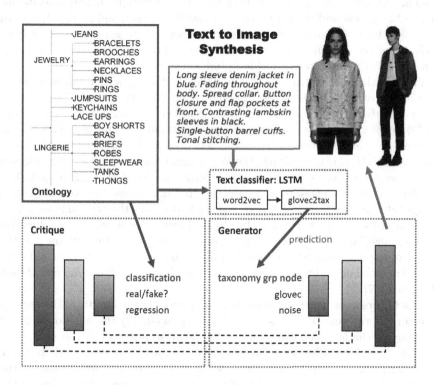

Fig. 1. Diagram of the proposed O-GAN with the ontology information flow depicted in red. The *classification output* of the critique uses the ontology as an additional classification task to improve model training. The *regression output* of the critique uses the word embeddings as an additional regression task to create a better match between the generated images and their description. The generator is conditioned on the *taxonomy group nodes* provided by the *text classifier*, as well as the word embedding of the description. Generated images together with the corresponding textual description are shown on the top right.

---

[1] https://fashion-gen.com/.

2 Ontology Generative Adversarial Networks

Recently a new training methodology for GANs, namely progressive growing of GANs (PGAN [5]) was proposed to improve the variation, stability and quality for image generation. PGAN starts the training process with low-resolution images and gradually increases the resolution, and extends its neural network model's architecture in order to fit the input image's resolution. However, this model is not capable of dealing with label ontologies. We propose *Ontology Generative Adversarial Networks (O-GANs)* that adopts the progressive training scheme of PGANs to cope with the challenges of high-resolution image generation, and further incorporate label ontology and text conditioning into the model. A block-diagram of our proposed method is provided in Fig. 1. As can be seen, our model consists of 3 modules: (1) the critique network to distinguish between real and generated images, classify images, and predict the word embedding from an image, (2) the generator network to generate images from conditioning vectors, and (3) the text classifier to predict *taxonomy group node*[2] from the text description of a fashion product.

The Critique is constructed having three outputs each minimizing one of three respective objectives: (1) for distinguishing between real and fake images, (2) for classifying images into their *group node* in the taxonomy of the labels, and (3) for predicting the word vector representations from images. This setting allows for parameter sharing between the networks used with different objectives. The first objective, is to minimize the Wasserstein distance [2] between real and fake (generated) images in the critique using the *discrimination output*. The second objective, is a *classification* objective that minimizes the categorical cross entropy between the *taxonomy group node* provided with the fashion images, and the *classification output*. Finally the third objective, is a *regression* objective that minimizes an L_2 loss between the *regression output* of the critique, and the average of the word embeddings, created using the description provided with the image. For word embedding, we use the *Global Vectors for Word Representation (glovectors)* [7].

The Generator uses three concatenated vectors as the input: (1) the average glovectors of the text as the *text conditioning input*, (2) the one-hot encoded taxonomy as the *label conditioning input*, and (3) a uniform random noise.

The Text-Classifier uses the sequences of glovectors extracted from the text descriptions as input, and predicts the best matching *group node* in the taxonomy. We use a bidirectional LSTM [4] trained on one-hot encoding of the taxonomy labels. This prediction is used during generation time as the label conditioning input to the generator.

---

[2] As the labels in the FashionGen dataset have two levels of the fashion taxonomy, a one-hot encoding for categories and subcategories are used as the *taxonomy group node*.

3 Evaluation Measures

We use three evaluation measures to compare the proposed method with the PGAN baseline. We evaluate the quality of the generated images from different models based on the Fréchet Inception Distance (FID) [3] and Inception Score (IS) [9] (computed on 1k generated and 1k real images). The Inception model was trained on ImageNet [1]. To evaluate the quality of the conditioning, we report the cross-entropy between the conditioning labels and the probability of the labels for generated images, estimated via the *classification output* in the critique. In addition, we report the L_2 distance between the average glovectors extracted from the conditioning text, and an estimated glovector via the regression output in the critique. The regression output also shares its weights in the critique, but has a separate output layer trained to minimize the L_2 distance between images and their corresponding glovector. Hence, the L_2 distance measures the similarity between the generated image and the text used for the conditioning.

4 Results and Discussion

In Table 1, we provide evaluation metrics on the Fashion-Gen dataset. Generated images of the implemented O-GAN are shown in Fig. 1 together with the corresponding textual description. We used a modified PGAN as a baseline, in which a *classification output* is added in addition to PGAN's discriminator. To demonstrate the importance of the ontologies, we only trained this baseline with category labels which has no information about the ontology of the finer-level classes. Additionally the generator was modified to use the label as conditioning vector. As shown in Fig. 2, we can see that the O-GAN achieves lower L_2 distance to the conditioning glovector, compared to PGAN. This suggests that the images generated by O-GAN have higher similarity to the conditioning text compared to the PGAN. It can also be seen that a lower cross-entropy between the conditioning labels and the classification output can be achieved which demonstrates better ability in label conditioning in O-GAN compared to PGAN. As can be seen in Table 1 and Fig. 2, the proposed method outperforms the PGAN in all cases of the reported evaluations.

Table 1. Image quality evaluation results. For Inception Score (IS) higher values and for Fréchet Inception Distance (FID) lower values are better.

	IS (1k)↑	FID (1k)↓
Real images	5.03 ± 0.36	0
Vanilla PGAN	4.54 ± 0.28	33.80
Proposed O-GAN	**4.81 ± 0.61**	**31.14**

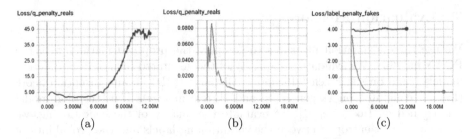

Fig. 2. Comparison of critique losses for the modified PGAN (blue) and the proposed O-GAN (magenta). (a) L_2 distance between the condition glovector and the estimated glovector from images in PGAN by the regression output in different epochs. (b) L_2 distance between the condition glovector and the estimated glovector from images in O-GAN by the regression output in different epochs. (c) Cross-Entropy between label condition and the probabilities of the *classification output* from the generated images in PGAN and O-GAN in different epochs. (Color figure online)

5   Conclusion

In this paper, we proposed a new Text-Conditional GAN, the O-GAN, capable of incorporating label ontology in conditioning, for generating high-resolution images only from a text description. We demonstrated the ability of O-GAN in incorporating ontologies in the generative process and showed how it improves the performance of both conditioning and quality of the generated images.

References

1. Deng, J., et al.: ImageNet: a large-scale hierarchical image database. In: CVPR 2009, pp. 248–255 (2009)
2. Gulrajani, I., et al.: Improved training of Wasserstein GANs. In: NIPS 2017, pp. 5769–5779 (2017)
3. Heusel, M., et al.: GANs trained by a two time-scale update rule converge to a local Nash equilibrium. In: NIPS 2017, pp. 6629–6640 (2017)
4. Hochreiter, S., Schmidhuber, J.: Long short-term memory. Neural Comput. **9**(8), 1735–1780 (1997)
5. Karras, T., et al.: Progressive growing of GANs for improved quality, stability, and variation. In: ICLR 2018 (2018)
6. Kuang, Z., et al.: Ontology-driven hierarchical deep learning for fashion recognition. In: MIPR 2018, pp. 19–24 (2018)
7. Pennington, J., et al.: Glove: global vectors for word representation. In: EMNLP 2014, pp. 1532–1543 (2014)
8. Rostamzadeh, N., et al.: Fashion-Gen: the generative fashion dataset and challenge. arXiv:1806.08317 (2018)
9. Salimans, T., et al.: Improved techniques for training GANs. In: NIPS 2016, pp. 2234–2242 (2016)

TIR

Introducing the Notion of 'Contrast' Features for Language Technology

Marina Santini[1](✉), Benjamin Danielsson[2], and Arne Jönsson[2,3]

[1] Division ICT-RISE SICS East, RISE Research Institutes of Sweden,
Stockholm, Sweden
marina.santini@ri.se
[2] Department of Computer and Information Science, Linköping University,
Linköping, Sweden
benda425@student.liu.se, arne.jonsson@liu.se
[3] Division ICT-RISE SICS East, RISE Research Institutes of Sweden,
Linköping, Sweden

Abstract. In this paper, we explore whether there exist 'contrast' features that help recognize if a text variety is a genre or a domain. We carry out our experiments on the text varieties that are included in the Swedish national corpus, called Stockholm-Umeå Corpus or SUC, and build several text classification models based on *text complexity features, grammatical features, bag-of-words features* and *word embeddings*. Results show that text complexity features and grammatical features systematically perform better on genres rather than on domains. This indicates that these features can be used as 'contrast' features because, when in doubt about the nature of a text category, they help bring it to light.

Keywords: Genre · Domain · Supervised classification · Features

1 Introduction

Finding a neat divide across text varieties is a difficult exercise. In the experiments presented in this paper, we focus on two text varieties, i.e. genre and domain. We observe that domains have a topical nature (e.g. Medicine and Sport), while genres have a communicative and textual nature (e.g. academic articles, health records, prescriptions or patient leaflets). A domain normally includes documents belonging to several genres; for instance, genres such as patient leaflets, articles and prescriptions are commonly used in the medical domain. Conversely, individual genres may serve several domains (like academic articles) or can be peculiar to a single domain (like health records). Sometimes, genre and domain are conflated together in domain adaptation, a field associated with parsing, machine translation, text classification and other NLP tasks. However, researchers have recently pointed out that mixing up text varieties has a detrimental effect on the final performance. For instance, some researchers (e.g.

© Springer Nature Switzerland AG 2019
G. Anderst-Kotsis et al. (Eds.): DEXA 2019 Workshops, CCIS 1062, pp. 189–198, 2019.
https://doi.org/10.1007/978-3-030-27684-3_24

[14] and [15]) analyse how topic-based and genre-based categories affect statistical machine translation, and observe that translation quality improves when using separate genre-optimized systems rather than a one-size-fits-all genre-agnostic system.

Since it is not straightforward to sort out genre from domain and vice versa, we investigate whether it is possible to decide automatically if a text variety is a genre or a domain. For instance, is "hobby" a genre or a domain? What about "editorial" or "interview"? In this paper, we explore whether there exist 'contrast' features that help recognize if a text category is a **genre** or a **domain**. By 'contrast' features we refer to those features that consistently perform well (or badly) only on one text variety. We explore the text categories that are included in the Swedish national corpus, called Stockholm-Umeå Corpus or SUC. We build several supervised text classification models based on several feature sets, namely *text complexity features, grammatical features, bag-of-words (BoW) features* and *word embeddings*.

2   Working Definitions

Before setting out any computational explorations across the text varieties of the SUC, we would like to provide working definitions that help understand the core difference between genre and domain.

Domain is a subject field. Generally speaking, domain refers to the shared general topic of a group of texts. For instance, "Fashion", "Leisure", "Business", "Sport", "Medicine" or "Education" are examples of broad domains. In text classification, domains are normally represented by topical features, such as content words (i.e. open class features like nouns and verbs) and specialized terms (such as "anaemia" in Medicine, or "foul" in Sport).

The concept of **genre** is more abstract than domain. It characterizes text varieties on the basis of conventionalized textual patterns. For instance, an "academic paper" obeys to textual conventions that differ from the textual conventions of a "tweet"; similarly, a "letter" complies to communicative conventions that are different from the conventions of an "interview". "Academic papers", "tweets", "letters" and "interviews" are examples of genres. Genre conventions usually affect the organization of the document (its rhetorical structure and composition), the length of the text, vocabulary richness, as well as syntax and morphology (e.g. passive forms vs. active forms). In text classification, genres are often represented by features such as Parts-of-Speech (POS) tags, character n-grams, POS n-grams, syntactic tags and function words.

In this complex scenario, how can we assess computationally whether a text category is a genre or domain? We propose using 'contrast' features, as explained in the next sections.

3   Previous Work

The diversified nature of the text varieties of the SUC has proved to be problematic when the corpus has been used in automatic genre classification experiments.

For instance, it has been observed that the SUC's text categories have a dissimilar nature since some of the SUC's 'genres' are in fact subject-oriented [13]. Additionally, researchers have recommended to work with a more complete and uniform set of SUC genres [13]. Previous SUC genre classification models [13] were based on four types of features, i.e. "words, parts-of-speech, parts-of-speech plus subcategories (differentiating between, e.g., personal and interrogative pronouns), and complete Parole word classifications (which include gender, tense, mood, degree, etc.)." Previous results show that grammatical features tend to perform slightly better than word frequencies when taking all the 9 SUC categories into account. The overall performance was rather modest though, showing wide variations across the 9 SUC's text categories.

Interestingly, a similar low performance was reported also when building genre classification models [10] based on discriminant analysis, 20 easy-to-extract grammatical features and 10 or 15 text categories of the Brown corpus [5]. The Brown corpus is the predecessor of the SUC and contains as well a mixtures of text varieties. More encouraging results were achieved when grouping the Brown corpus' text categories into two subgroups and four subgroups [10].

In situations like those described above [10, 13], one can surmize that there is probably not enough labeled training data in the datasets to get higher performance. It could also be that the learning algorithm is not well suited to the data, or that a lower error rate is simply not achievable because the predictor attributes available in the data are insufficient to predict the classes more accurately. However, we argue that this is not the case here, since more recent experiments where SUC categories have been sorted out into different varieties give a different picture. For instance, SUC text varieties have been interpreted as mixture of domains, genres, and miscellaneous categories [4]. The reordering of SUC text varieties was proposed as follows: six "proper" genres (i.e. Press Reportage (A), Press Editorial (B), Press Review (C), Biographies/Essays (G), Learning/Scientific Writing (J) and Imaginative Prose (K); two subject-based categories or domains (Skills/Trades/Hobbies (E) and Popular Lore (F); and a mixed category, i.e. Miscellaneous (H). The findings show that genre classification based on readability features and the whole SUC (9 classes) has a modest performance, while using readability features only for the subset of six 'proper genres' gives much better results.

Text complexity features (another way to refer to readability features) have also shown good performance on SUC's proper genres [9]. More precisely, PCA-based components extracted from text complexity features perform remarkably well on five proper genres, namely Press Reportage (A), Press Editorial (B), Press Review (C), Learning/Scientific Writing (J) and Imaginative Prose (K) [9].

In the experiments described below we offer a more systematic picture of how to use features to decide about the nature of text varieties.

4   The SUC: Corpus and Datasets

The SUC is a collection of Swedish texts (amounting to about one million words) and represents the Swedish language of the 1990's [6]. The SUC follows the general layout of the Brown corpus [5] and the LOB corpus [8], with 500 sample texts

Table 1. Set 1. Experiments with text complexity features and grammatical features.

Set 1	SUC text categories	Features	SMO	DI4jMlp
Exp1	9 SUC varieties (a_reportage_genre, b_editorial_genre, c_review_genre, e_hobby_domain, f_popular_lore_domain, g_bio_essay_genre, h_miscellaneous_mixed, j_scientific_writing_genre, k_imaginative_prose_genre); 1400 instances	115 complexity features	**0,596**	0,582
		65 components	0,567	0,572
		27 POS tags	0,507	0,526
		62 dependency tags	0,541	0,531
Exp2	5 SUC genres (a_reportage_genre, b_editorial_genre, c_review_genre, j_scientific_writing_genre, k_imaginative_prose_genre); 682 instances	115 complexity features	**0,831**	0,813
		64 components	0,829	0,811
		27 POS tags	0,786	0,773
		62 dependency tags	0,782	0,771
Exp3	4 SUC varieties (2 domains and 2 genres: e_hobby_domain, f_popular_lore_domain, j_scientific_writing_genre, k_imaginative_prose_genre); 402 instances	115 complexity features	**0,785**	0,766
		58 components	0,722	0,704
		27 POS tags	0,743	0,740
		62 dependency tags	0,715	0,711
Exp4	2 SUC genres (j_scientific_writing_genre, k_imaginative_prose_genre); 216 instances	115 complexity features	0,981	0,981
		51 components	0,972	0,949
		27 POS tags	**0,986**	0,981
		62 dependency tags	0,981	0,968
Exp5	2 SUC domains (e_hobby_domain, f_popular_lore_domain,); 186 instances	115 complexity features	0,720	**0,749**
		55 components	0,692	0,674
		27 POS tags	0,674	0,706
		Dependency tags	0,707	0,722

with a length of about 2,000 words each. The SUC text varieties are somehow fuzzy although they are collectively called "genres" by the corpus creators [6]. It is worth stressing that the SUC was created to represent the Swedish language as a whole, and not to represent different text varieties. Certainly, this corpus design affects text categorization models.

Technically speaking, the SUC is divided into 1040 bibliographically distinct text chunks, each assigned to so-called "genres" and "subgenres".

The source dataset containing linguistic features was created using the publicly available toolkit named TECST [3] and the text complexity analysis module called SCREAM [7]. The dataset contains 120 variables [2], thereof 115 linguistic features, three readability indices (LIX, OVIX and NR), and two descriptive variables (file name and the language variety). In these experiments, we only used 115 linguistic features.

The BoW dataset was created from the annotated SUC 2.0 corpus, which can be downloaded from SpråkBanken[1] (Gothenburg University).

---

[1] See https://spraakbanken.gu.se/eng/resources/corpus.

5   Experiments

The experiments in this study are all based on supervised machine learning, which implies that the classification models are trained and built on labelled data. In this case, the labelled data is the SUC's text categories. The general idea behind supervised classification is that once a model proves effective on a set of labelled data (i.e. the performance on the test data is good), then the model can be safely applied to unlabelled data. The requirement is that the unlabelled data on which a supervised classification model is going to be applied has a similar composition and distribution of the data on which the supervised model has been trained. The supervised paradigm differs from unsupervised classification (clustering) where classification models are trained and built on completely unlabelled data. In that case, a human intervention is needed to name and evaluate the resulting clusters.

In this section, we present three sets of experiments and a detailed report of their performance. The first set is based on text complexity- and grammatical features; the second set relies on BoW features; in the third set we present a comparative experiment based on function words and word embeddings. As mentioned above, we apply supervised machine learning and rely on two stable learning models, namely Support Vector Machines (SVM) and Multilayer Perceptron (MLP). We use off-the-shelf implementations of SVM and MLP to ensure full replicability of the experiments presented here. We run the experiments on the Weka workbench [16]. Weka's SVM implementation is called SMO and includes John Platt's sequential minimal optimization algorithm for training a support vector classifier[2]. Weka provides several implementations of MLP. We use the DI4jMlpClassifier, that is a wrapper for the DeepLearning4j library[3], to train a multi-layer perceptron. Both algorithms were run with standard parameters. Results are shown in Tables 1, 2 and 3. We compared the performance on the Weighted Averaged F-Measure (AvgF)[4] and applied 10-folds crossvalidation.

Set 1. The first set contains five experiments (see Table 1). In Experiment 1, models are created with all the 9 SUC text categories (1400 instances) using four different features sets, namely *115 text complexity features*, *PCA-based components*, *POS tags* and *dependency tags*. The two algorithms (SMO and DI4JMlpClassifier), although they have a very different inductive bias, achieve a very similar performance. We observe however that the DI4JMlpClassifier is slower than SMO, which is extremely fast: SMO took no more than a few seconds on all datasets, while the DI4JMlpClassifier's average processing time was about a couple of minutes. In this set, we observe that the performance on the "domain" varieties (i.e. "hobby" and "popular lore") is quite poor. The majority

---

[2] See http://weka.sourceforge.net/doc.dev/weka/classifiers/functions/SMO.html.

[3] See https://deeplearning.cms.waikato.ac.nz/.

[4] Weighted Averaged F-Measure is the sum of all the classes F-measures, each weighted according to the number of instances with that particular class label. It is a more reliable metric than the harmonic F-measures (F1).

Table 2. Set 2. Experiments with bag-of-words features

Set 2	SUC text categories	BOW features	SMO	DI4jMlp
Exp1	9 SUC varieties (a_reportage_genre, b_editorial_genre, c_review_genre, e_hobby_domain, f_popular_lore_domain, g_bio_essay_genre, h_miscellaneous_mixed, j_scientific_writing_genre, k_imaginative_prose_genre); 1400 instances	Including stopwords	**0,767**	0,640
		Without stopwords	0,741	0,614
Exp2	5 SUC genres (a_reportage_genre, b_editorial_genre, c_review_genre, j_scientific_writing_genre, k_imaginative_prose_genre); 682 instances	Including stopwords	**0,903**	0,854
		Without stopwords	0,863	0,824
Exp3	4 SUC varieties (2 domains and 2 genres: e_hobby_domain, f_popular_lore_domain, j_scientific_writing_genre, k_imaginative_prose_genre); 402 instances	Including stopwords	**0,905**	0,828
		Without stopwords	0,880	0,792
Exp4	2 SUC genres (j_scientific_writing_genre, k_imaginative_prose_genre); 216 instances	Including stopwords	**0,991**	**0,991**
		Without stopwords	**0,991**	**0,991**
Exp5	2 SUC domains (e_hobby_domain, f_popular_lore_domain,); 186 instances	Including stopwords	**0,925**	0,858
		Without stopwords	0,892	0,842

of the texts labelled "hobby" were classified as "reportage" (65 instances). The same thing happened with the majority of the "popular lore" texts. One could surmize that this happens because "reportage" is the most populated class in the dataset. Interestingly, however, a very small class like "bio_essay", that contains two genres, is not attracted towards "reportage". Unsurprisingly, also the "miscellaneous" class has a high number of misclassified texts, notably 42 out of 145 miscellaneous texts have been classified as "reportage", and only 70 texst have been classified with the correct label.

In Experiment 2, we created models with only five single "proper genres". We ditched out the "bio-essay" class because it included two genres that are not necessarily close to each other and because the class was very little populated. All the instances labelled with text varieties other than the five genres were removed, and we ended up with a model built on 682 instances. We observe that the performance increases dramatically (up to AvgF = 0.831 when using 115 complexity features) if compared to the models in Experiment 1.

In Experiment 3, a balanced dataset was created with two domains and two genres. The best performance is achieved by SMO based on 115 complexity features. We observe that the performance on the two domains (and especially on "popular lore") is definitely lower than the performance on the two genres. This decline could be interpreted as a sign that text complexity features are not representative of topic-based varieties.

In Experiment 4, we created models with two genres only, and we observe that the performance soars up dramatically (up to AvgF = 0.986) when using only 27 POS tags and SMO.

In Experiment 5, we created models with two domains only. The performance is definitely lower, reaching its peak with 115 complexity features in combination with the DI4jMlpClassifier (AvgF = 0.749). Again, the "popular lore" domain suffers from many misclassifications. We interpret the moderate performance on these two subject-based classes as the effect of the inadequacy of the features to represent the nature of domain-based text varieties. From this set of experiments, it appears that text complexity- and grammatical features are genre-revealing features, and their performance on topic-based categories is lower.

Set 2. Also the second set contains five experiments, but this time models are built with BoW features. Two BoW datasets were used: one including stopwords, and the other one without stopwords. Stopwords are normally removed when classifying topic-based categories, while they are helpful for genre classification [14]. A breakdown of the results is shown in Table 2. In Exp1, the best performance is achieved by SMO in combination with BoW+stopwords (AvgF = 0.767). This overall performance is much higher than in Exp1, Set 1. BoW features perform much better on the domains and on the miscellaneous class. The inclusion of stopwords helps genre classification. This is good news for the classification task in itself, but less so for the distinction between genre and domain, that is important in some other NLP tasks, as mentioned earlier. In Exp2, Exp3, Exp4 and Exp5, the best performance is achieved by SMO in combination with BoW features including stopwords. AvgF is very high in all cases (always greater than 0.90), reaching the peak (AvgF = 0.991) on two proper genres.

Set 3. Set 3 contains only a single experiment. In this set of experiments, we compare the contrastive power of two very different feature sets – function words and word embeddings – on the 9 SUC text categories. Function words (a.k.a. stopwords) are grammatical closed classes. They can be used in the form of POS tags (like here) or as word frequencies. They are light-weight features that can be successfully used for genre detection [14], although they are less powerful than other features [12]. Here we used 15 POS tags that represent function words. It took a few seconds to create the models.

Word embeddings are one of the most popular representation of document vocabulary to date. They can be used for many tasks (e.g. sentiment analysis, text classification, etc.). Word embeddings are capable of capturing the context of a word in a document, as well as semantic and syntactic similarity. Here we use Word2Vec word embeddings [11], extracted with the unsupervised Weka filter called DI4jStringToWordVec in combination with the MI4jMLPClassifier (SMO cannot handle word embeddings). Training a neural networks-based classifier is time consuming. After laborious parameters' tuning, we used a configuration based on three convolutional layers, a global pooling layer, a dense layer and an output layer. It took 5 days to run this configuration with 10-folds crossvalidation. Results (shown in Table 3) are definitely modest on the SUC dataset.

Table 3. Function words (15 POS tags) vs Word2Vec word embeddings

Set 3	SUC text categories	Features	SMO	DI4jMlp
Exp	9 SUC varieties (a_reportage_genre, b_editorial_genre, c_review_genre,	Function Words	0,371	**0,448**
	e_hobby_domain, f_popular_lore_domain, g_bio_essay_genre, h_miscellaneous_mixed, j_scientific_writing_genre, k_imaginative_prose_genre); 1400 instances	Word Embeddings	n/a	0.340

Discussion. In the first set of experiments, we explored whether text complexity features (taken individually or aggregated in PCA-based components) and grammatical features have enough contrastive power to disentangle genres and domains. It turns out that these features are more representative of genres than domains and mixed classes since they perform consistently better on genre classes, as neatly shown in all the five experiments in Table 1. It appears that they can be safely used as 'contrast' features. In most cases, the best performance is achieved with 115 text complexity features in combination with SMO. In the second set of experiments, BoW features perform equally well on genres and on domains. The best performance is achieved with BoW features including stopwords. All in all, the classification performance of BoW features on the five experiments is higher than the performance based on text complexity- and grammatical features. However, it is unclear whether the BoW models are generalizable. We speculate that these models somehow overfit the corpus (although we applied 10-folds crossvalidation). For the purpose of our investigation, we observe that BoW features have no or little contrastive power, since their behaviour is rather indistinct across genres and domains. In the third set of experiments, we compared the performance of two very different feature, namely function words and word embeddings. Both feature sets are quite weak. All in all, function words perform better in combination with SMO. We observe that although the overall performance is quite modest, function words are very effective with the "Reportage" genre and the "Imaginative prose" genre, where the number of misclassifications is very limited. Word embeddings are a thorny kind of feature. It has been pointed out that "[d]espite the popularity of word embeddings, the precise way by which they acquire semantic relations between words remain unclear" [1]. Their performance on the SUC is definitely modest.

6   Conclusion and Future Work

In this paper, we argued that text varieties have a diversified nature. We limited our empirical investigation to two text varieties, namely genre and domain. The core of our investigation was the quest of 'contrast' features that could automatically distinguish between genre classes and domain classes. We explored the contrastive power of several feature sets, and reached the conclusion that text complexity features and grammatical features are more suitable as 'contrast' features than BoW features. In particular, text complexity features perform consistently better on genres than on domains. This means that these features can help out when in doubt about the nature of text varieties. Function words and word embeddings seem less suitable as 'contrast' features. A valuable by-product of the empirical study presented here is a comprehensive overview of the performance of different feature sets on the text varieties included in the SUC.

Future work includes the exploration of additional 'contrast' features as well as the application of this approach to other corpora containing mixed text varieties (e.g. the Brown corpus and the British National Corpus).

Acknowledgements. This research was supported by E-care@home, a "SIDUS – Strong Distributed Research Environment" project, funded by the Swedish Knowledge Foundation [kk-stiftelsen, Diarienr: 20140217]. Project website: http://ecareathome.se/

References

1. Altszyler, E., Sigman, M., Slezak, D.F.: Corpus specificity in LSA and word2vec: the role of out-of-domain documents. arXiv preprint arXiv:1712.10054 (2017)
2. Falkenjack, J., Heimann Mühlenbock, K., Jönsson, A.: Features indicating readability in Swedish text. In: Proceedings of the 19th Nordic Conference of Computational Linguistics (NoDaLiDa-2013), No. 085 in NEALT Proceedings Series 16, Oslo, Norway, pp. 27–40. Linköping University Electronic Press (2013)
3. Falkenjack, J., Rennes, E., Fahlborg, D., Johansson, V., Jönsson, A.: Services for text simplification and analysis. In: Proceedings of the 21st Nordic Conference on Computational Linguistics, pp. 309–313 (2017)
4. Falkenjack, J., Santini, M., Jönsson, A.: An exploratory study on genre classification using readability features. In: Proceedings of the Sixth Swedish Language Technology Conference (SLTC 2016), Umeå, Sweden (2016)
5. Francis, W.N., Kucera, H.: Brown Corpus Manual: Manual of Information to Accompany a Standard Corpus of Present-Day Edited American English for Use with Digital Computers. Brown University, Providence (1979)
6. Gustafson-Capková, S., Hartmann, B.: Manual of the Stockholm Umeå Corpus version 2.0. Stockholm University (2006)
7. Heimann Mühlenbock, K.: I see what you mean. Assessing readability for specific target groups. Dissertation, Språkbanken, Department of Swedish, University of Gothenburg (2013). http://hdl.handle.net/2077/32472
8. Johansson, S., Leech, G.N., Goodluck, H.: Manual of information to accompany the Lancaster-Oslo/Bergen Corpus of British English, for use with digital computer. University of Oslo, Department of English (1978)

9. Jönsson, S., Rennes, E., Falkenjack, J., Jönsson, A.: A component based approach to measuring text complexity. In: Proceedings of the Seventh Swedish Language Technology Conference 2018 (SLTC-2018) (2018)
10. Karlgren, J., Cutting, D.: Recognizing text genres with simple metrics using discriminant analysis. In: Proceedings of the 15th Conference on Computational Linguistics, vol. 2, pp. 1071–1075. Association for Computational Linguistics (1994)
11. Mikolov, T., Sutskever, I., Chen, K., Corrado, G.S., Dean, J.: Distributed representations of words and phrases and their compositionality. In: Advances in Neural Information Processing Systems, pp. 3111–3119 (2013)
12. Santini, M.: Automatic Identification of Genre in Web Pages: A New Perspective. LAP Lambert Academic Publishing, Saarbrücken (2011)
13. Wastholm, P., Kusma, A., Megyesi, B.: Using linguistic data for genre classification. In: Proceedings of the Swedish Artificial Intelligence and Learning Systems Event, SAIS-SSLS (2005)
14. Van der Wees, M., Bisazza, A., Monz, C.: Evaluation of machine translation performance across multiple genres and languages. In: Proceedings of the Eleventh International Conference on Language Resources and Evaluation (LREC-2018) (2018)
15. Van der Wees, M., Bisazza, A., Weerkamp, W., Monz, C.: What's in a domain? Analyzing genre and topic differences in statistical machine translation. In: Proceedings of the 53rd Annual Meeting of the Association for Computational Linguistics and the 7th International Joint Conference on Natural Language Processing (Volume 2: Short Papers), vol. 2, pp. 560–566 (2015)
16. Witten, I.H., Frank, E., Hall, M.A., Pal, C.J.: Data Mining: Practical Machine Learning Tools and Techniques. Morgan Kaufmann, Cambridge (2016)

LineIT: Similarity Search and Recommendation Tool for Photo Lineup Assembling

Ladislav Peška[1,3(✉)] and Hana Trojanová[2]

[1] Department of Software Engineering, Faculty of Mathematics and Physics, Charles University, Prague, Czech Republic
peska@ksi.mff.cuni.cz
[2] Department of Psychology, Faculty of Arts, Charles University, Prague, Czech Republic
[3] T-Labs, Faculty of Informatics, ELTE Budapest, Budapest, Hungary

Abstract. In this paper we focus on the problem of assembling unbiased photo lineups. Photo lineups are an important method in the identification process for the prosecution and possible conviction of suspects. Incorrect lineup assemblies have led to the false identification and conviction of innocent persons. One of the significant errors which can occur is the lack of lineup fairness, i.e., that the suspect significantly differs from other candidates.

Despite the importance of the task, few tools are available to assist police technicians in creating lineups. Furthermore, these tools mostly focus on fair lineup administration and provide only a little support in the candidate selection process. In our work, we first summarize key personalization and information retrieval (IR) challenges of such systems and propose an IR/Personalization model that addresses them. Afterwards, we describe a LineIT tool that instantiate this model and aims to support police technicians in assembling unbiased lineups.

Keywords: Information retrieval · Recommender systems · Photo lineups

1 Introduction

Eyewitness testimony and identification often play a significant role in criminal proceedings. A very important part of such identification is the lineup, i.e. the eyewitness's selection of the offender among other lineup members. However, there are cases of incorrect identifications in lineups which led to the conviction of innocent suspects.

One of the principal recommendations for inhibiting errors in the identification is to assemble fair (unbiased) lineups, i.e., to choose lineup candidates who are sufficiently similar to the suspect w.r.t. a given description [8]. Imagine a witness, who, based on some external circumstances (e.g., distance from the

© Springer Nature Switzerland AG 2019
G. Anderst-Kotsis et al. (Eds.): DEXA 2019 Workshops, CCIS 1062, pp. 199–209, 2019.
https://doi.org/10.1007/978-3-030-27684-3_25

perpetrator, insufficient lighting, elapsed time), does not clearly remember the offender. However, if the lineup is biased (e.g., the suspect is the only member with matching approximate age, ethnicity, body build, or similar), the witness could select the suspect only based on these general features without actually recognising him/her [4].

Fair lineups minimize the probability of such *"best-available"* identifications (also denoted as *relative judgement*) both by decreasing the rate of "best-available" selections (i.e., in unbiased lineups, the witness may realize his/her uncertainty and will not identify anyone) and by decreasing the probability of its correctness (i.e., other lineup candidates, for whom we know that they are not the offenders, appears similar enough to the suspect and therefore, the witness could select them as the "best-available" choice instead).

Although there are various research projects e.g., [7] as well as commerce activities, e.g., elineup.org, aiming to the process of eyewitness identification, these tools mostly focus on the lineup administration problems and provide only a limited support for the lineup assembling task. To the best of our knowledge, there is no related research on systems supporting intelligent lineup assembling, visual similarity search, or recommending lineup candidates, except for our own previous work [10–12].

On the other hand, there are several similar challenges that inspired us when we were materializing our vision. The idea of utilizing unstructured visual data to support information retrieval (IR) processes gain a novel impulse with the dawn of the deep convolutional networks (DCNN) [6]. A common DCNN's task is image classification, ranging from general images' class hierarchy [13] up to the identification of individual persons [9] from their photos. It has also been shown that semantic descriptors corresponding with network's activation can be treated as a high-dimensional feature vector of the respective object and transferred for another learning task [2]. We utilized such approach while deriving visual descriptors of lineup candidates from the VGG-Face network [9].

Recommending lineup candidates may seem somewhat related to the problem of session-based recommendations, due to the expected drift in user's preferences[1]. In session-based recommendations, however, the underlying assumption is that user's information needs may gradually or suddenly change also during the session, which favors, e.g., recurrent neural networks [5]. In the lineup assembling, however, user's task and his/her information needs are largely pre-determined by the current lineup or suspect and are supposedly stable throughout the lineup processing. Therefore, instead of a fully temporal model, we merely utilize per-lineup feedback in our retrieval models.

In the rest of the paper, we summarize previous works into the list of key IR and personalization challenges specific for the considered task and propose an IR/Personalization model addressing these challenges. Afterwards, we describe and evaluate the LineIT tool, a prototype application aiming to assist police

---

[1] The user's task and therefore expected recommendations may significantly differ between each session. Similarly, for every new lineup, there is a different suspect, which can be considered as a topic drift as well.

technicians in assembling unbiased photo lineups. The LineIT tool exploits both visual descriptors of candidates' photos as well as their content-based (CB) attributes, such as age, nationality or appearance characteristics. The LineIT GUI features (i) a (multiple) query-by-example (QBE) search engine with CB pre-filtering and the capability to target local features and (ii) a candidate recommender system (RS) combining both visual and CB recommendations via fuzzy D'Hondt's elections scheme.

2 IR Models for Lineup Assembling

While considering personalization and information retrieval challenges in the lineup assembling task, we took into account the following observations.

First, in our initial work [10], two item-based recommendation models were utilized: the cosine similarity of weighted CB attributes and the cosine similarity of latent visual descriptors from the VGG-Face network. Although the recommendations based on visual descriptors significantly outperformed the CB ones, both methods received positive feedback and the intersection of proposed candidates was very small. Supposedly, both methods focus on a different subset of aspects, which together form a human-perceived similarity of persons.

This assumption was validated in a follow-up paper [11] via off-line evaluation, which further demonstrated that a substantial improvement of CB model's performance can be achieved, if the weights of individual CB features are personalized. The personalization was done via minimizing the weighted distance between suspects and candidates selected by the current police technician. However, such method did not improve the results of visual recommendations.

The questionnaire among participating police technicians [10] also revealed the need for *lineup uniformity*, i.e., not just the perceived distance of candidates towards the suspect, but all intra-lineup distances should be considered.

In [12], we compared the quality of lineups created by several domain experts vs. automatically assembled lineups aiming to maximize the lineup's uniformity. The evaluation showed that both approaches are comparable w.r.t. lineup fairness, yet the intersection of lineup members was very small. Furthermore, the variance of per-lineup results was rather high and often, while the one method provided reasonable results, the others failed. Therefore, we may conclude that we need a "human-in-the-loop" solution, where automated content processing tools (e.g., recommender systems) may significantly reduce the processing time and increase lineup fairness.

The high variability of per-lineup results indicates that also the difficulty of processing each lineup varies greatly, i.e., simple selections from recommended candidates might not be sufficient for more difficult lineups. Therefore, users should have some options to explicitly query the candidates. A combination of CB filtering and visual QBE search are currently standard tools for visual retrieval systems. The extension of QBE to multiple examples is a natural consequence of the need for lineup uniformity and works well in some other domains [15]. However, the analysis of the quality of assembled lineups [12] revealed also

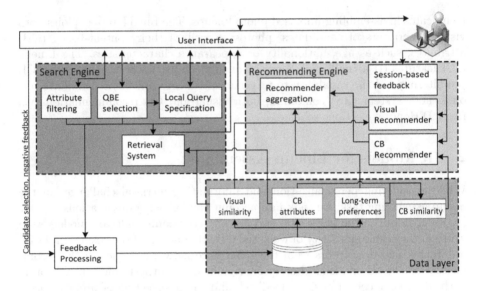

Fig. 1. The proposed Retrieval model for the intelligent lineup assembling task.

the need for a local search, i.e., searching for specific visual features, such as a particular haircut, mouth shape, skin defect etc. This is important in cases, where such features were explicitly noted by a witness, or the suspect possess some rare but determining appearance characteristics. A portion of local search needs may be instantiated via attribute-based filtering, but it is also possible to focus the visual similarity of a particular region of a candidate's photo [14].

The personalization of the lineup assembling task is somewhat controversial as the adaptation towards police technician's preferences may itself create a bias in assembled lineups. Therefore, as a minimum, the effect of personalization on resulting lineup fairness should be monitored. On the other hand, perceived similarity may vary among human beings and users (similarly as in any other task) will probably reject recommendations that are relevant according to some generic metric, but irrelevant according to their own one[2].

The lineup assembling is a highly session (i.e., lineup) dependent task and all retrieval models should primarily focus on the user's feedback related to the current session. However, there are also several opportunities for learning long-term user's preferences. First, we have already shown that the importance of content-based attributes could be tuned to better fit a user's needs [11]. Second, while selecting lineup candidates, users exhibited different propensity towards Visual or CB similarity of persons, so both approaches may be combined in a personalized way to derive final recommendations. Lastly, the propensity of users towards lineup uniformity may be learned similarly as the propensity of users towards diversified recommendations [1]. The schema of the proposed IR/Personalization model is depicted on Fig. 1. The model operates with CB and Visual similar-

---

[2] I.e., we do not aim on "educating" users about the "correct" similarity metric.

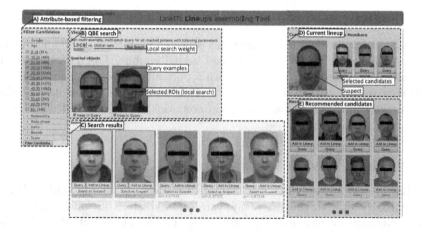

Fig. 2. Screenshot of the LineIT Tool with a partially assembled lineup.

ity and employs QBE and attribute-based search together with Visual and CB recommendations. Several extensions to the current model are possible, e.g., employing the negative user feedback. Both explicit (disapproving a search result or a recommendation) and implicit (ignoring proposed objects or GUI features) feedback may provide valuable information for model tuning. Also, although the collaborative filtering is not directly applicable, the similarity of users may be explored. However, the low expected volume of users may be a challenge.

3   LineIT Tool

The LineIT tool is a prototype application aiming to evaluate the practical applicability of the proposed IR/Personalization model. As such, it has some limitations, e.g., a static dataset, interim visualization, the lack of authentication etc., which should be addressed before a final deployment. LineIT is implemented as an open-source Python/Flask based web server, client-side scripting (JavaScript) powers highly responsive user interface, while recommendations and event responses are operated via AJAX calls. Main LineIT components are depicted on Fig. 2: (a) attribute-based filtering and (b) QBE definition are processed together as a single query and (c) its results are depicted in lower central area. Note the optional selection of multiple queried examples and utilizing local search via specifying regions of interest (ROIs) within the query examples. The lineup definition component (Fig. 2d) manages the current suspect and already selected candidates and the right lower section (Fig. 2e) displays candidates recommended according to the current lineup and long-term user preferences (based on past user's feedback).

3.1 Dataset

In this demo application, the dataset was collected from the wanted and missing persons application of the Police of the Czech Republic. The dataset contains in total 4,336 males and 739 women with a portrait photo, nationality, age and some appearance characteristics. Similar characteristics as well as dataset sizes are generally available in the real-world scenarios. For a detailed description, please refer to [15].

3.2 Similarity Models

The LineIT tool incorporates two similarity models: visual and content-based. CB similarity utilizes the list of appearance characteristics and employs the cosine similarity of weighted CB attributes. The weights of attributes' values are learned from the past user feedback.

Visual descriptors corresponds with the activation of the last fully connected layer (FC8) of the VGG-Face network [9] initialized with the person's photo. Visual similarity of persons c_1 and c_2 is defined as the cosine similarity of their visual descriptors $\mathcal{V}_{c_1}$ and $\mathcal{V}_{c_2}$.

$$sim_G(c_1, c_2) = cosSim(\mathcal{V}_{c_1}, \mathcal{V}_{c_2}) \tag{1}$$

In order to allow users to search for local features, we split candidates' photos into several regions of interest (ROIs)[3], which can be selected as relevant by the user. The local similarity of persons c_1 and c_2 w.r.t. selected ROIs L is defined as

$$sim_L(c_1, c_2, L) = \sum_{i \in L} \frac{\max_{j \in s(i)} cosSim(\mathcal{V}_{c_1}^i, \mathcal{V}_{c_2}^j)}{|L|}, \tag{2}$$

where $\mathcal{V}_c^i$ denotes a visual descriptor corresponding to the i-th ROI of a person c and $s(i)$ denotes all regions similar to the region i. This definition is based on the observation that although potentially interesting visual features (e.g., eyes, ears, mouth) maintain similar positions within the image, this may vary and therefore it might be useful to incorporate also the neighboring ROIs of other persons.

3.3 Search Engine

Upon receiving a query, the search engine operates in two phases. In the first phase, it creates a list of acceptable candidates $\mathcal{C}_a \subseteq \mathcal{C}$ according to the hard constraints defined on CB attributes' values.

In the second phase, the list of acceptable candidates is ranked according to its similarity with the selected query examples $\mathcal{C}_q = \{(c_i, L_i)\}$. Let w_L be the

---

[3] Currently, a fixed 3×4 grid with a 10% padding is used, but other variants, e.g., based on the detection of key facial artefacts are plausible.

overall importance of the local search[4]. Then a similarity of a candidate c_j to the query $\mathcal{C}_q$ is defined as

$$sim(\mathcal{C}_q, c_j) = (1 - w_L) \times \sum_{\forall c_i \in \mathcal{C}_q} \frac{sim_G(c_i, c_j)}{|\mathcal{C}_q|} + w_L \times \sum_{\forall (c_i, L_i) \in \mathcal{C}_q} \frac{sim_L(c_i, c_j, L_i)}{|\mathcal{C}_q|}$$

(3)

Top-k candidates with the highest similarity to the query $\mathcal{C}_q$ are reported to the user.

3.4 Recommending Engine

The recommending module operates asynchronously based on the observed user feedback. For now, changes in the assembled lineup, i.e., adding/removing candidates or suspect, selection of recommended candidate and changes in attribute-based filtering affects the recommendations. While changes in the current state of the assembled lineup is considered as a proxy towards user's short-term preferences (i.e., upon the change, a new list of recommended candidates is generated), the selection of recommended candidates and applied attribute-based filters are considered as sources of user's long-term preferences.

While processing the recommending requests, the following principles are maintained. First, the uniformity of the lineup is maintained, i.e., recommended candidates should be similar to the suspect as well as to all already selected candidates [10]. Second, both content-based and visual-based similarity represents (up to some extent) an approximation of human-perceived similarity and therefore, candidates from both recommenders should be appropriately represented. However, both models produce highly diverse results, so the weighting-based aggregations would inevitably prefer one of the recommenders. Therefore, other aggregation mechanisms should be considered.

Lineup Uniformity in Individual Recommenders. The problem of lineup uniformity is addressed as follows. In the Visual recommender, the rating r_i^V of a candidate c_i is defined as the weighted average of candidates' similarities with the current lineup members $\mathcal{C}_L$:

$$r_i^V = \sum_{(c_j, w_j) \in \mathcal{C}_L} \frac{w_j * sim_G(c_i, c_j)}{W},$$

(4)

where W is the sum of all weights w_j. Currently, the suspect is prioritized, while the weight of other lineup members is uniform.

In CB recommender, the candidate's rating r_i^{CB} is defined analogically, however, the long-term user's preferences on attribute values, $\mathcal{W}_A$, are considered.

$$r_i^{CB} = \sum_{(c_j, w_j) \in \mathcal{C}_L} \frac{w_j * cosSim(\mathcal{W}_A \circ \mathcal{A}_{ci}, \mathcal{W}_A \circ \mathcal{A}_{cj})}{W}$$

(5)

---

[4] This weight is currently set explicitly via a slider, however, it can be also incorporated in user's long-term preferences in the future.

$\mathcal{A}_c$ denotes the vector of candidate's content-based attributes and $\circ$ denotes element-wise multiplication.

Fuzzy D'Hondt's Recommender Aggregation. While considering the alternatives to aggregate recommenders, we realized that our scenario shares some properties with political elections. For this task, D'Hondt's election method [3] is used to transform the volume of votes to the volume of mandates, while maintaining a fair ranking of candidates and the proportionality of votes to mandates.

For parties $p_i \in \mathcal{P}$, an ordered list of candidates per party, $C_i \subset \mathcal{C}$, and a total volume of votes per party, v_i, the method iteratively selects the next candidate from the party, which currently has the most accountable votes. Accountable votes, a_i, are defined as $v_i/(k_i + 1)$, where k_i is the volume of already selected candidates from party p_i. The limitation of the original method is that it cannot account the fuzzy membership of candidates to the parties, which may happen in our case. Therefore, we proposed its fuzzy extension as follows: lists of all candidates and their per-party relevance score, $\{(c_j, r_j^i) : c_j \in \mathcal{C}, p_i \in \mathcal{P}\}$, are maintained. In each step, the relevance of each candidate is calculated as a weighted sum of accountable votes

$$r_j = \sum_{\forall p_i \in \mathcal{P}} a_i * r_j^i \qquad (6)$$

The candidate with the highest relevance, say c_m, is selected. Then, fuzzy counters and per-party accountable votes are updated as follows $\forall p_i \in \mathcal{P} : k_i = k_i + r_m^i; a_i = v_i/(k_i + 1)$.

In our use-case, we represent recommenders as parties, top-k recommended persons correspond with the party's candidates and their relevance scores w.r.t. each party are normalized into a unit vector to prevent bias. The volume of votes per party is a hyperparameter of the model, which is learned from user's long-term preferences.

User's Long-Term Preferences. Two types of user's preferences are considered in LineIT: the propensity towards CB or Visual recommendations and the importance of individual CB attributes. The CB vs. Visual tradeoff is learned via applying a gradient descend step maximizing the candidate's relevance (6) upon its selection. The initial values are set according to our previous experiments [10].

The importance of attributes' values, $\mathcal{W}_A$, is initialized via a non-negative least squares optimization[5] trained on the difference of the suspects' and lineup members' CB descriptors. The importance is then fine-tuned for the current user, based on the values he/she selected in the attribute-based filtering, i.e., the relative importance of an attribute increases every time it was selected by the user.

---

[5] docs.scipy.org/doc/scipy/reference/generated/scipy.optimize.nnls.html.

4   Usage Evaluation

In order to evaluate individual features of the proposed system, we conducted a modest user study and asked 11 users to assemble lineups of several "suspects" selected from the database. The implicit feedback describing the utilization of systems component was collected during the process of lineups assembling with the aim to reveal, which of the GUI components are extensively utilized and which lead to acceptable solutions.

Results showed that in the majority of cases (206 out of 371), users selected candidates offered by the recommender system. In the remaining 165 cases, a search GUI candidates were used. While analysing the search queries in general, some CB conditions were specified in 65%, QBE in 87% and local search in 52% of queries. In total 72% of queries were multimodal, i.e., incorporating at least two of the previously mentioned modalities. This corresponds with our previous results supporting the major role of QBE search, but also shows that other proposed components are important for the users. The importance of QBE search is further supported by the fact that although users sometimes experimented with attribute-only queries (13% of queries), these only rarely lead to the suitable results (7% of selected candidates). On the other hand, searching by multiple examples were only rarely utilized by the users (in 7% of queries). We would probably attribute this to the lower comprehensibility of this setting, because recommendations based on multiple examples worked well according to the users' selections. However, we plan to address this phenomenon in more details in our future work.

5   Conclusions and Future Work

In this paper, we focused on the problem of assembling unbiased photo lineups. In the first part, we summarized the current knowledge and transformed it into the proposal of IR/Personalization model for the intelligent assembling of lineups. In the second part, we presented and evaluated the LineIT tool, a prototype tool for assembling unbiased lineups that instantiates this model. The main aim of the LineIT tool is to provide a reasonable test bed for evaluating various features of the proposed model, but it is also a first step towards a deployment-ready lineup assembling software.

Future Work. Both the IR model and the LineIT tool should be extensively evaluated in the future. First results suggest usability of the proposed concepts, but additional experiments, e.g., focused on the quality of proposed lineups or long-term usage patterns are needed. Furthermore, there were certain design choices, e.g., the selection of similarity metrics or election-based ensemble, which deserves a more complete justification and experimental evaluation. Apart from evaluation of its own properties, LineIT provides an opportunity to evaluate capabilities of police technicians, while facing various edge cases in lineup assembling, e.g., different suspect's ethnicity etc. This direction of research may end up both with the list of recommendations for praxis as well as updated IR models that better comply with these edge cases.

Acknowledgements. This work was supported by the grants GAUK-232217, GACR-19-22071Y and 20460-3/2018/FEKUTSTRAT Higher Education Excellence Program 2018. Some resources are available online:
LineIT demo: http://herkules.ms.mff.cuni.cz/lineit_v2,
Source codes: https://github.com/lpeska/LineIT_v2.

References

1. Di Noia, T., Rosati, J., Tomeo, P., Sciascio, E.D.: Adaptive multi-attribute diversity for recommender systems. Inf. Sci. **382**(C), 234–253 (2017)
2. Donahue, J., et al.: DeCAF: a deep convolutional activation feature for generic visual recognition. CoRR abs/1310.1531 (2013). http://arxiv.org/abs/1310.1531
3. Gallagher, M.: Proportionality, disproportionality and electoral systems. Electoral Stud. **10**(1), 33–51 (1991)
4. Greene, E., Heilbrun, K.: Wrightsman's Psychology and the Legal System. Cengage Learning, Belmont (2013)
5. Hidasi, B., Karatzoglou, A.: Recurrent neural networks with top-k gains for session-based recommendations. In: Proceedings of the 27th ACM International Conference on Information and Knowledge Management, CIKM 2018, pp. 843–852. ACM (2018)
6. Krizhevsky, A., Sutskever, I., Hinton, G.E.: Imagenet classification with deep convolutional neural networks. In: Advances in Neural Information Processing Systems 25, pp. 1097–1105. Curran Associates, Inc. (2012)
7. MacLin, O.H., Zimmerman, L.A., Malpass, R.S.: Pc_eyewitness and the sequential superiority effect: computer-based lineup administration. Law Hum Behav. **29**(3), 303–321 (2005)
8. Mansour, J., Beaudry, J., Kalmet, N., Bertrand, M.I., Lindsay, R.C.L.: Evaluating lineup fairness: variations across methods and measures. Law Hum Behav. **41**, 103 (2016)
9. Parkhi, O.M., Vedaldi, A., Zisserman, A.: Deep face recognition. In: British Machine Vision Conference (2015)
10. Peska, L., Trojanova, H.: Towards recommender systems for police photo lineup. In: Proceedings of the 2nd Workshop on Deep Learning for Recommender Systems, DLRS 2017, pp. 19–23. ACM (2017)
11. Peska, L., Trojanova, H.: Personalized recommendations in police photo lineup assembling task. In: Proceedings of the 18th Conference Information Technologies - Applications and Theory (ITAT 2018), No. 2203 in CEUR-WS (2018)
12. Peska, L., Trojanova, H.: Towards similarity models in police photo lineup assembling tasks. In: Marchand-Maillet, S., Silva, Y.N., Chávez, E. (eds.) SISAP 2018. LNCS, vol. 11223, pp. 217–225. Springer, Cham (2018). https://doi.org/10.1007/978-3-030-02224-2_17
13. Russakovsky, O., et al.: ImageNet large scale visual recognition challenge. Int. J. Comput. Vis. (IJCV) **115**(3), 211–252 (2015)

14. Skopal, T., Peška, L., Grošup, T.: Interactive product search based on global and local visual-semantic features. In: Marchand-Maillet, S., Silva, Y.N., Chávez, E. (eds.) SISAP 2018. LNCS, vol. 11223, pp. 87–95. Springer, Cham (2018). https:// doi.org/10.1007/978-3-030-02224-2_7
15. Skopal, T., Peška, L., Kovalčík, G., Grosup, T., Lokoč, J.: Product exploration based on latent visual attributes. In: Proceedings of the 2017 ACM on Conference on Information and Knowledge Management, CIKM 2017, pp. 2531–2534. ACM (2017)

Advanced Behavioral Analyses Using Inferred Social Networks: A Vision

Irena Holubová[1] , Martin Svoboda[1(✉)] , Tomáš Skopal[1] ,
David Bernhauer[1,2] , and Ladislav Peška[1]

[1] Faculty of Mathematics and Physics, Charles University, Prague, Czech Republic
{holubova,svoboda,skopal,peska}@ksi.mff.cuni.cz
[2] Faculty of Information Technology, Czech Technical University in Prague,
Prague, Czech Republic
bernhdav@fit.cvut.cz

Abstract. The success of many businesses is based on a thorough knowledge of their clients. There exists a number of supervised as well as unsupervised data mining or other approaches that allow to analyze data about clients, their behavior or environment. In our ongoing project focusing primarily on bank clients, we propose an innovative strategy that will overcome shortcomings of the existing methods. From a given set of user activities, we infer their social network in order to analyze user relationships and behavior. For this purpose, not just the traditional direct facts are incorporated, but also relationships inferred using similarity measures and statistical approaches, with both possibly limited measures of reliability and validity in time. Such networks would enable analyses of client characteristics from a new perspective and could provide otherwise impossible insights. However, there are several research and technical challenges making the outlined pursuit novel, complex and challenging as we outline in this vision paper.

Keywords: Inferred social networks · Similarity · Behavioral analysis

1 Introduction

The behavior, environment, and characteristics of clients form a significant source of information for various businesses in order to increase their revenues, detect potential problems, prevent unwanted situations, or at least suppress their negative impact. During the first stage of our ongoing project, we are focusing on clients of banks and other similar financial institutions providing basic banking products such as deposit accounts, payment cards or cash loans. However, analogous principles and challenges can be observed in other areas too, such as, e.g., landline or mobile phone operators, electricity or gas suppliers, or other.

This work was supported in part by the Technology Agency of the Czech Republic (TAČR) project number TH03010276 and by Czech Science Foundation (GAČR) project number 19-01641S.

G. Anderst-Kotsis et al. (Eds.): DEXA 2019 Workshops, CCIS 1062, pp. 210–219, 2019.
https://doi.org/10.1007/978-3-030-27684-3_26

In the financial sector, banks primarily evaluate various behavioral or credit scores of both potential and existing clients in order to classify them as fraudulent or legitimate, with various risk levels involved. There are basically three main issues that banks essentially need to solve: (1) fraud detection, (2) debt services default, and (3) customer churn. In addition, various life situations (e.g., finishing studies, getting married, the birth of a child, promotion, loss of a job, retirement, etc.) can be interesting for banks, so that they can offer appropriate financial products to relevant clients, and offer them effectively and in the right moment.

For the purpose of the analysis of bank clients, there are several suitable sources of information. The main options involve: (1) socio-demographic or other information provided by the clients themselves, (2) transactional history or other financial interaction of a client with the bank, and (3) publicly available third-party information that can be linked with the internal client data. In theory, especially the last option seems to provide valuable information so that the identified concerns of banks can be tackled. In practice, however, there are various limitations, technical as well as legal (data and privacy protection or other regulations), which might not permit to use this approach in its full extent.

There exists a number of data mining approaches that try to solve some of the indicated problems [31]. Most of the solutions use some kind of a supervised method [12,16,21], where we can predict the future behavior of clients based on a learning data set where the corresponding targets were well known or identified manually. Unsupervised approaches [23] use characteristics of clients in order to group them into clusters on the basis of their mutual similarity, trying to maximize the difference between them.

In this paper, we propose a basic concept and principles of a novel solution inspired by both of these approaches based on the idea of analyzing a social network of clients. Of course, also in this context the idea of utilizing social networks was not only already proposed [2] (e.g., for fraud detection [14,19,27] or peer-to-peer lending [17]), but this need is also confirmed by bank representatives themselves. However, in our approach, we go much further. First, we do not assume an existence of such a network (which is a strong, often unrealistic assumption), but based on the available characteristics and history of financial transactions we construct an *inferred* social network. For this purpose, we exploit not only direct facts when describing client relationships, but also features inferred from the available data indirectly using various similarity methods and statistical approaches, all that with additional measures of *reliability* and time-limited *validity* of the information.

Contemporary solutions deployed in banks are often based on techniques such as traditional data processing and querying (projection, selection, and aggregation), integration of externally available information, processing of strings (substrings, n-grams, common subsequences or regular expressions), natural language processing, logistic regression, decision trees, or neural networks. These can help us to solve partial problems such as household, salary, or installment detection. For this purpose, both client and transactional data are exploited: counterparty account numbers are matched to well-known collection accounts (e.g., big suppliers, tax offices, etc.), text comments or other payment symbols accompanying bank transfers analyzed, or descriptions of merchants associated with card payment transactions categorized. All in all, these techniques primarily use rather obvious, direct and only factual kind of information. Having implemented our vision, a new set of possibilities opens, simply because we will be able to apply behavioral patterns observed on clients who just tend to be similar (i.e., clients who would otherwise not be related to each other at all, let alone because they have absolutely no direct or obvious relationships), and simulate propagation of such information, bank decisions and offers, their consequences and impact through the network. For example, two clients can be considered similar, just because they have similar distribution of monthly card payments based on different *Merchant Category Codes* (MCC), money withdrawals carried out abroad, or unexpectedly dwindled overall account balance normalized to the average achieved salary.

As indicated so far, such envisioned networks will permit the banks to view and analyze clients, their characteristics, behavior, and mutual relationships from a new perspective, and provide otherwise impossible insights. However, the outlined vision cannot be attained straightforwardly, since it poses several technical as well as research challenges. In this paper we (1) describe the first results of our project, i.e., the process of inference of the social network from information in bank transactions (see Sect. 2) and (2) envision the open problems and challenges of its analysis and exploitation (not only) for the financial sector (see Sect. 3). We believe that these two contributions will trigger a new research direction for the information retrieval community applicable in many other domains.

2   Inferred Social Network

In our first use case from the financial sector, we assume the possession of the following input (anonymized) data:

1. client characteristics (e.g., name, addresses, education, etc.),
2. history of financial transactions (e.g., credit/debit card payments, payments with cash back service, cash withdrawals, transfers, permanent transfer orders, regular or overdue loan installments, etc.), and, eventually,
3. third-party information related to the non-person entities from the previous two data sets (e.g., shops, institutions, etc.).

In all the cases, only data belonging to a particular monitored time interval $I = [t_{start}, t_{end}]$ is covered. In the following text, we provide a more precise definition of the target social network and the process of its inference from the input data.

2.1   Social Network Graph

We construct the social network of bank clients at two different levels of granularity, representing a dual view of the same problem from different perspectives:

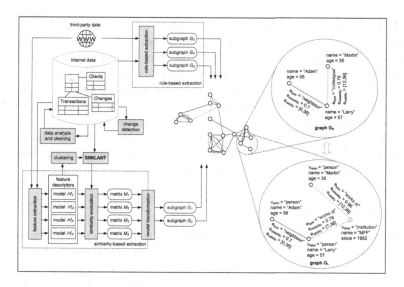

Fig. 1. The process of inference of a social network

(1) *high-level* view, where the social network only consists of vertices for bank clients, and (2) *low-level* view, where there are vertices also for different kinds of real-world entities (e.g., institutions, companies, etc.). Both the views are depicted in Fig. 1 as graphs G_H and G_L on the right.

High-Level Network. We define a *high-level network* to be a multigraph $G_H = (V_H, E_H)$, where the set of vertices V_H represents individual bank clients and the set of edges E_H relationships among the clients. Each client $v \in V_H$ has a set of properties $\{p_1, p_2, ..., p_n\}$, each of which is modeled as a tuple with the following components:

- name p_{name} (e.g., *age*),
- value p_{value} (e.g., *58*),
- reliability $p_{reliability} \in [0, 1]$ (e.g., *0.85* for information certain for 85%), and
- validity $p_{validity} = [t_{start}, t_{end}]$ representing the time interval of validity of the information.

Each edge $e \in E_H$ has the following components:

- relationship type e_{type} (e.g., *colleague*),
- reliability $e_{reliability} \in [0, 1]$, and
- validity $e_{validity} = [t_{start}, t_{end}]$.

The high-level view G_H has the advantage of having a simple data structure to work with. It enables to perform general analyses without the need to distinguish between types of vertices and to know and understand technical details of the reality. For example, we are 85% sure that two clients work in the same company, but we do not (need to) know which company.

Low-Level Network. In case we are interested in more details (e.g., we want to know particulars about a given company), we can use the low-level view G_L, where we also work with vertices for different real-world entities, possibly enriched by third-party publicly available information about them.

Formally, we model this *low-level network* as a multigraph $G_L = (V_L, E_L)$, where the set of vertices V_L involves V_H plus vertices for new kinds of entities, each one of them newly associated also with a vertex type v_{label}, as well as properties $\{p_1, p_2, ..., p_n\}$ with the unchanged structure as in the original high-level network G_H. The edges in E_L represent relationships among clients and institutions, locations, etc., also having e_{type}, $e_{reliability}$, and $e_{validity}$.

2.2 Inference Process

As usual with real-world data sets, the data first need to be pre-processed using a *data analysis and cleaning* module. In this step, we can already identify information whose reliability is <1 due to lower data quality. However, this feature will mainly be adjusted by the inference process. Similarly, from the input data, we already know the basic time intervals of validity to be further refined.

The process of inference of graph G_H (or G_L) consists of two orthogonal approaches which we denote as *rule-based* extraction and *similarity-based* extraction (see Fig. 1 on the left).

Rule-Based Extraction. The more straightforward inference path is based on the idea of the definition of a set of domain-specific rules, each of which defines a way of constructing a particular type (or types) of relationships in the target graph G_H (or G_L). So far, we distinguish the following classes:

- *factual relationships* directly present in the input data (e.g., a married couple),
- *aggregated relationships* derived from the input data using various information aggregations (e.g., co-workers defined as people working concurrently in the same company), and
- *transitive relationships* derived from the input data on the basis of statistically significant amount of occurrences of specific events (e.g., people living in the same household defined as people with the same home address, frequent mutual money transfer, frequent payments in similar shops, etc.). In this case usually $e_{reliability} < 1$.

Similarity-Based Extraction. The process of data extraction in this inference path is more complex and, to the best of our knowledge, also unique in the given context, so we describe it in more detail. We first define a set of similarity models $\mathcal{M}_1, \mathcal{M}_2, \ldots, \mathcal{M}_k$, each of which expresses a particular similarity of users defined by certain selected *features*. These features can be of various types, such as based on single fixed values, time-varying series, accumulated values (e.g., per week or month), etc., or their combinations. With regards to the monitored time interval

I, we can focus on the whole I, or just its part (e.g., since the moment a client entered a contract with the bank).

Within a model $\mathcal{M}_l$, values of such features for a given particular client c are represented using a descriptor d_c^l; mostly a vector or time series where individual coordinates/elements represent values in the selected features (e.g., a normalized overall volume of performed debit card payment transactions, converted into the base currency, all that separately for every month of interval I). Having a set of descriptors for all the clients, standard approaches can be exploited to calculate similarities of the clients using the selected metrics, configured weights, or other parameters [30].

For each model $\mathcal{M}_l$, this step thus outputs a dense square matrix M_l, where for each two vertices v_i and $v_j \in V_H$ (or V_L), the measure of their mutual similarity $sim_l(v_i, v_j) \in [0, 1]$ with regards to the corresponding descriptors is stored at $M_l[i, j]$. The efficient matrix computation[1] could be implemented by similarity self-join on V_H (or V_L), e.g., using the Hadoop MapReduce algorithms [6].

Next, the set of dense matrices $M_1, M_2, ..., M_k$ is transformed to a set of sparse matrices $M_1', M_2', ..., M_m'$ (k and m do not necessarily need to be equal), each representing one type of relationships to be added into the target graph, each particular edge e with the value of reliability $e_{reliability}$ corresponding to $M_l'[i, j]$. The "sparsing" transformation can involve application of similarity thresholds, combination of multiple similarity results (e.g., using a weighted sum), clustering of clients into categories, etc. In general, we want to restrict the result just to the most interesting information with a reasonable size.

To summarize, the similarity-based extraction is a method for indirect detection of relationships between clients (or entities, in general). While the rule-based extraction can reach $e_{reliability} = 1$ and well-defined semantics of the relationships, there is usually a limited amount of data available. The similarity-based extraction, on the other hand, provides weaker semantics and $e_{reliability} < 1$ of relationships (in fact, it is correlated with the similarity scores), but it still allows to identify relationships in situations where the rule-based extraction is not applicable (due to, e.g., a small amount of data, a small number of direct relationships, etc.).

SIMILANT. For the analysis of promising similarity models to be used in the similarity-based extraction, we have designed and implemented an analytics tool called SIMILANT. It enables effective browsing of data based on the chosen similarity measure. It consists of three independent parts: clustering, visualization, and browser. During the extraction process, we create a complete weighted graph (i.e., the dense similarity matrix M_l), which is, however, hard to interpret. Instead of its visualization, we visualize only a small number of related clusters. The browser part of SIMILANT provides simple analytics of similarity models at different levels of detail (e.g., different numbers of these clusters). Their semantics can be further validated using various *targets* (i.e., labeled ground truth), if available.

---

[1] Instead of fully materialized matrices approximations can be used.

Dynamics of the Input Data. An important feature of the input data set is its dynamics. One aspect is that new data continuously appear. Another one is the event-driven nature of the problem domain. For example, there are various life situations (e.g., the loss of a job or a notable increase in salary) highly influencing the financial behavior of bank clients and, consequently, the structure of the graph representing their financial behavior. In order to prevent skewed data, a *change detection* module determines points of such changes using a set of predefined domain-specific rules combined with statistical analyses. When such a change connected with a given client is identified at time t_{change}, instead of working with the entire interval $I = [t_{start}, t_{end}]$, we can technically split a given client behavior into two sub-intervals $I_1 = [t_{start}, t_{change}]$ and $I_2 = [t_{change}, t_{end}]$ – before and after the change – and study a given client within the two (or more) intervals.

2.3   Social Network Analysis

Having the inferred (high-level or low-level) social network, in the next phase of our project, we will focus on its thorough analysis. Currently, there exist several verified approaches for social network analysis which primarily need to be utilized for validity and reliability of the information. In addition, since the inferred network is not built by people themselves, quite probably its features will not correspond to features of traditional social networks. As a consequence, verified approaches may need to be optimized, modified, or even identified as inapplicable for inferred networks.

In general, our main near-future target areas are as follows:

- *Visualization*: We will focus on the visualization of the inferred network with filters related to both validity and reliability. In addition, we need to visualize the dynamics of the graph.
- *Structural Analysis*: We will utilize traditional methods [14], like, e.g., analysis of density, centrality, structural holes, clustering coefficient, communities, etc. Due to the specifics of inferred networks, we assume that the characteristics will not correspond to usual observations for social networks [24].
- *Information Propagation*: Inspired by sociology, we will deal with how information is propagated [8] in inferred networks, primarily using *independent cascade* or *linear threshold* models, both progressive stochastic diffusion models.

These steps will lead to the main target: to identify the most relevant information for the financial sector in order to solve the three key issues and optimize customer services.

3   Challenges

The envisioned idea of inferring a social network from information describing people and their interactions can be used in many other domains. Regardless of the domain, such inferred social networks will probably have different features and thus will require novel approaches for their processing and analysis. We believe that this idea opens a new, challenging, and highly practically applicable research area for the information retrieval community. Besides the previous list of target research areas of our project, we summarize other related challenges of inferred networks as follows:

- *Big Graph Data*: Even a small bank can have hundreds of thousands of clients and hundreds of transactions per a client and month and the inferred social network can be very large. Approaches, such as *network embedding* [28], that allow us to compact the information necessary to describe its structure can increase the efficiency of its processing. Even if this should lead to approximations. Other challenges are observed in big graphs with high-degree vertices, randomness in graph structure or other irregularities, different types of edges, or in graphs with dynamic changes, causing sharding, data locality or load balancing issues [20,25,29]. Data quality issues, although studied for decades, also require attention in the context of Big Data processing [3,13,15].
- *Time-Varying Data*: Various time aspects of features and behavior of people are natural. Hence, management of data with a temporal dimension has already been addressed in a number of fields, from relatively old *temporal databases* [9,22] for relational data to newer *time-varying networks* [11] for which there exists an extensive amount of specific approaches [4], including dedicated data stores [5]. Inferred social networks have the time aspect too, however, its processing needs to be further adjusted with regards to their other specific features, such as (time-varying) reliability.
- *Feature Selection*: The similarity-based extraction path brings a challenge in exclusion of redundant or irrelevant descriptors whose number can be extremely large. While the *supervised* approaches (which require additional information) are well explored [1,7,18], the *unsupervised* ones are more suitable in the considered context, but also more challenging and thus less represented. For example, several criteria and unsupervised approaches are discussed in [10], whereas paper [26] proposes how to measure the validity of feature selection objectively. Since the target inferred social network cannot be extremely dense, the choice of the most relevant data is a crucial issue in general.

References

1. Ang, J.C., Mirzal, A., Haron, H., Hamed, H.N.A.: Supervised, unsupervised, and semi-supervised feature selection: a review on gene selection. IEEE/ACM Trans. Comput. Biol. Bioinform. **13**(5), 971–989 (2016). https://doi.org/10.1109/TCBB.2015.2478454
2. Baesens, B., Vlasselaer, V.V., Verbeke, W.: Fraud Analytics Using Descriptive, Predictive, and Social Network Techniques: A Guide to Data Science for Fraud Detection, 1st edn. Wiley, Hoboken (2015)

3. Cai, L., Zhu, Y.: The challenges of data quality and data quality assessment in the big data era. Data Sci. J. **14**(2), 1–10 (2015). https://doi.org/10.5334/dsj-2015-002
4. Casteigts, A., Flocchini, P., Quattrociocchi, W., Santoro, N.: Time-varying graphs and dynamic networks. In: Frey, H., Li, X., Ruehrup, S. (eds.) ADHOC-NOW 2011. LNCS, vol. 6811, pp. 346–359. Springer, Heidelberg (2011). https://doi.org/10.1007/978-3-642-22450-8_27
5. Cattuto, C., Quaggiotto, M., Panisson, A., Averbuch, A.: Time-varying social networks in a graph database: a Neo4j use case. In: First International Workshop on Graph Data Management Experiences and Systems, GRADES 2013, pp. 11:1–11:6. ACM, New York (2013). https://doi.org/10.1145/2484425.2484442
6. Čech, P., Maroušek, J., Lokoč, J., Silva, Y.N., Starks, J.: Comparing mapreduce-based k-NN similarity joins on hadoop for high-dimensional data. In: Cong, G., Peng, W.-C., Zhang, W.E., Li, C., Sun, A. (eds.) ADMA 2017. LNCS (LNAI), vol. 10604, pp. 63–75. Springer, Cham (2017). https://doi.org/10.1007/978-3-319-69179-4_5
7. Chandrashekar, G., Sahin, F.: A survey on feature selection methods. Comput. Electr. Eng. **40**(1), 16–28 (2014). https://doi.org/10.1016/j.compeleceng.2013.11.024
8. Chen, W., Lakshmanan, L.V., Castillo, C.: Information and Influence Propagation in Social Networks. Synthesis Lectures on Data Management, vol. 5, no. 4, pp. 1–177 (2013). https://doi.org/10.2200/S00527ED1V01Y201308DTM037
9. Date, C.J., Darwen, H., Lorentzos, N.A.: Temporal Data and the Relational Model. Elsevier, Amsterdam (2002)
10. Dy, J.G., Brodley, C.: Feature subset selection and order identification for unsupervised learning. In: Proceedings of the Seventeenth International Conference on Machine Learning, October 2000
11. Holme, P., Saramäki, J.: Temporal networks. Phys. Rep. **519**(3), 97–125 (2012)
12. Islam, S.R., Eberle, W., Ghafoor, S.K.: Mining bad credit card accounts from OLAP and OLTP. CoRR abs/1807.00819 (2018). http://arxiv.org/abs/1807.00819
13. Katal, A., Wazid, M., Goudar, R.H.: Big data: issues, challenges, tools and good practices. In: 2013 Sixth International Conference on Contemporary Computing (IC3), pp. 404–409, August 2013. https://doi.org/10.1109/IC3.2013.6612229
14. Kirchner, C., Gade, J.: Implementing social network analysis for fraud prevention (2011)
15. Kwon, O., Lee, N., Shin, B.: Data quality management, data usage experience and acquisition intention of big data analytics. Int. J. Inf. Manag. **34**(3), 387–394 (2014). https://doi.org/10.1016/j.ijinfomgt.2014.02.002
16. Lessmann, S., Baesens, B., Seow, H., Thomas, L.C.: Benchmarking state-of-the-art classification algorithms for credit scoring: an update of research. Eur. J. Oper. Res. **247**(1), 124–136 (2015). https://doi.org/10.1016/j.ejor.2015.05.030
17. Lin, M., Prabhala, N.R., Viswanathan, S.: Judging borrowers by the company they keep: friendship networks and information asymmetry in online peer-to-peer lending. Manag. Sci. **59**(1), 17–35 (2013). https://doi.org/10.1287/mnsc.1120.1560
18. Liu, H., Yu, L.: Toward integrating feature selection algorithms for classification and clustering. IEEE Trans. Knowl. Data Eng. **4**, 491–502 (2005)
19. Lookman, S., Nurcan, S.: A framework for occupational fraud detection by social network analysis. In: Proceedings of the CAiSE 2015 Forum at the 27th International Conference on Advanced Information Systems Engineering co-located with (CAiSE 2015), Stockholm, Sweden, 10 June 2015, pp. 221–228 (2015). http://ceur-ws.org/Vol-1367/paper-29.pdf

20. Nai, L., Xia, Y., Tanase, I.G., Kim, H., Lin, C.: GraphBIG: understanding graph computing in the context of industrial solutions. In: Proceedings of the International Conference for High Performance Computing, Networking, Storage and Analysis, SC 2015, pp. 1–12, November 2015. https://doi.org/10.1145/2807591.2807626

21. Ngai, E.W.T., Hu, Y., Wong, Y.H., Chen, Y., Sun, X.: The application of data mining techniques in financial fraud detection: a classification framework and an academic review of literature. Decis. Support Syst. 50(3), 559–569 (2011). https://doi.org/10.1016/j.dss.2010.08.006

22. Ozsoyoglu, G., Snodgrass, R.T.: Temporal and real-time databases: a survey. IEEE Trans. Knowl. Data Eng. 7(4), 513–532 (1995). https://doi.org/10.1109/69.404027

23. Quah, J.T.S., Sriganesh, M.: Real-time credit card fraud detection using computational intelligence. Expert Syst. Appl. 35(4), 1721–1732 (2008). https://doi.org/10.1016/j.eswa.2007.08.093

24. Santoro, N., Quattrociocchi, W., Flocchini, P., Casteigts, A., Amblard, F.: Time-varying graphs and social network analysis: temporal indicators and metrics. In: 3rd AISB Social Networks and Multiagent Systems Symposium (SNAMAS), United Kingdom, pp. 32–38, May 2011. https://hal.archives-ouvertes.fr/hal-00854313

25. Singh, D.K., Patgiri, R.: Big graph: tools, techniques, issues, challenges and future directions. In: Sixth International Conference on Advances in Computing and Information Technology (ACITY 2016), pp. 119–128 (2016)

26. Tang, J., Alelyani, S., Liu, H.: Feature selection for classification: a review. In: Data Classification: Algorithms and Applications (2014)

27. Vlasselaer, V.V., et al.: APATE: a novel approach for automated credit card transaction fraud detection using network-based extensions. Decis. Support Syst. 75, 38–48 (2015). https://doi.org/10.1016/j.dss.2015.04.013

28. Wang, D., Cui, P., Zhu, W.: Structural deep network embedding. In: Proceedings of the 22nd ACM SIGKDD International Conference on Knowledge Discovery and Data Mining, pp. 1225–1234. ACM (2016)

29. Xia, Y., et al.: Graph analytics and storage. In: 2014 IEEE International Conference on Big Data (Big Data), pp. 942–951, October 2014. https://doi.org/10.1109/BigData.2014.7004326

30. Zezula, P., Amato, G., Dohnal, V., Batko, M.: Similarity Search - The Metric Space Approach. Advances in Database Systems, vol. 32. Kluwer, Dordrecht (2006). https://doi.org/10.1007/0-387-29151-2

31. Zhou, J.: Data mining for individual consumer credit default prediction under e-commence context: a comparative study. In: Proceedings of the International Conference on Information Systems - Transforming Society with Digital Innovation, ICIS 2017, Seoul, South Korea, 10–13 December 2017 (2017)

Author Index

Printed in the United States
By Bookmasters